Chemistry Experiments:
For Advanced & Honors Programs

James Signorelli

Order this book online at www.trafford.com
or email orders@trafford.com

Most Trafford titles are also available at major online book retailers.

Printed in the United States of America.

ISBN: 978-1-4907-4643-2 (sc)
ISBN: 978-1-4907-4644-9 (e)

Library of Congress Control Number: 2014916347

Trafford rev. 09/17/2014

www.trafford.com
North America & international
toll-free: 1 888 232 4444 (USA & Canada)
fax: 812 355 4082

Contents

Introduction...ix
LAB #1 Determination of the density..1
LAB #2 Investigating Avogadro's First Law (N= SM/MM)...........................7
LAB #3 Paper Chromatography of Ink..11
LAB #4 Inverse Square Law...15
LAB #5 Determining Spectral Lines for Hydrogen......................................19
LAB #6 Decomposition of Sodium Bicarbonate...25
LAB #7 Formula of a Hydrate..31
LAB #8 Lead Iodide Lab...34
LAB #9 Calculating the Molar Mass of Two Gases.....................................37
LAB #10 Chemical Reaction Rates...45
LAB #11 Ideal Gas Law – Expected Volume of a Gas.................................47
LAB #12 Enthalpy change for a double replacement reaction.....................52
LAB #12 Enthalpy change for a double replacement reaction.....................58
LAB #13 Determination of the Specific Heat Capacity of three pure metal samples 62
LAB #14 Melting Point behavior of a pure substance.................................66
LAB #15 Building a simple voltaic cell...71
LAB #16 Determination of the solubility of a compound in 100 ml of distilled water, at a constant temperature...77
LAB #17 Combustion of methyl alcohol...79
LAB #18 Calculating Absolute Zero...82
LAB #19 Titration of an acid solution..86
LAB #20 Single Replacement Reaction of Iron and Copper Sulfate............92
LAB #21 Observing the reaction rate of metals with water........................96
LAB #22 Esterification...98
LAB #23 The Hydrolysis of an Ester..103
LAB #24 Decomposition of Sucrose..106
LAB #25 Acetylation of Salicylic Acid: "Aspirin" production.....................108
LAB #26 AP level Synthesis of Bakelite...111
LAB #27 The preparation of a polymer, Thiokol, a synthetic rubber...........113
LAB #28 Determination of the diameter of a wire using density data..........115
LAB #29 Measuring the Surface Tension of a Liquid..................................117
LAB #30 Analytical Chemistry...120
LAB #31 Formula of a precipitate (Advanced Placement version)..............123
LAB #32 Laser Crystallography...128
LAB #33 Silly Slime..130
LAB #34 Properties of Carbon Dioxide gas...131
LAB #35 Properties of pure Hydrogen gas...133
LAB #36 Properties of pure Oxygen gas..135

Gifted and talented students, and any student interested in pursuing a science major in college needs a rigorous program to prepare them, while they are still in high school. This book utilizes a format where the application of several disciplines: science, math, and language arts principles are mandated. Each lab concludes with either an essay or a detailed analysis of what happened and why it happened. This format is based on the expectations of joining a university program, or becoming an industrial science professional.

The ideal student lab report would be written in a lab research notebook, and then the essay or final analysis is done on a word processor, to allow for repeat editing and corrections. The research notebook has all graph pages a title section, and a place for the students and their assistants to sign and witness that exercise. The basic mechanics of the lab report: title, purpose, procedure, diagrams, data table, math & calculations, observations, and graphs are hand written into the book. The conclusion is done on a word processor (MS-Word) which allows the instructor to guide the student in writing and editing a complete essay, using MLA format. When the final copy is completed, the essay is printed and inserted into the lab notebook for grading.

At the end of the term, the student has all of their labs in one place for future reference. These lab notebooks can be obtained for as little as $ 3.00 per book. This is money well spent. In our district, the Board of Education buys the books for each student. The B.O.E. sees these books as expendable, (but necessary), materials for all science and engineering instruction.

Introduction

At the honors and advanced placement levels, chemistry requires several areas of rigor for total mastery by the student. The mathematics used in the chemistry program must be at a minimum skill level of first year algebra. The language arts skills must be at a level where MLA format for essay writing is the norm, and not the exception. Last, but not least, the master student must make clear and precise connections of the physical science skills, connecting the concepts of chemistry, physics, biology, and mathematics. Often, it is found, that physics and chemistry share the same principles and investigations. For this reason, chemistry is often called the Central Science.

The state departments of education are adopting what is called Next Generation Science Standards (NGSS). The proper use of language arts skills is mandated by these new regulations. The master student does not make one word or simple phrase answers. The advanced student is expected to write very clear, precise, and inclusive essays, covering all of the related topics of the concepts being investigated. This essay should be in a format called MLA (Modern Language Association). The English department at all high schools requires this level of writing and so should the science department. These state departments are also adopting Common Core Curriculum Standards (CCCS). The new regulations require a much higher level of rigor than in the past. To better prepare students for college, the states are mandating changes to science programs in the public schools which makes them more like prep schools.

This lab manual addresses these new regulations. It requires the chemistry student to perform lab experiments which, until a few years ago, were only done in college. The purpose of science is to discover the mysteries of the universe. If the student cannot write a clear explanation of what happened, with cause and effect relationships, then those concepts remain a mystery. Although it may be difficult, mastery of any topic becomes the reward for those who put forth the extra effort.

LAB #1

Determination of the density

Purpose:

To determine the purity of four metal samples using the procedure established by Archimedes 2300 years ago.

Diagram:

Draw or attach a photo of the equipment in set-up mode

Procedure #1:

. Weigh the four metal samples to four decimal places (+/- 0.0001 grams)

. Using a Vernier Caliper, determine the diameter and length of each sample.

. Calculate the density using Archimedes formula.

. Calculate the purity of each pure metal sample (Al, Fe, Cu) percent error is related to purity.

. Calculate the percent composition of Zinc and Copper in the Brass sample.

. **Calculate the Density and hypothesize the identity of the unknown given in class.**

Data:

	Al	Fe	Cu	Brass unknown
Mass of sample				
Diameter of sample				
Radius of sample				
Height of sample				
Volume of sample				
Actual density				
Expected density	2.70	7.86	8.96	8.45
% error				

Calculations:

Show all math required to complete the table, in standard format

. Calculate the density of each sample.

. Calculate the number of _**moles**_ of atoms in the three (pure) elemental samples using N = SM / MM, where SM is the standard density of that metal. **(mass of 1 cm³)**

. Calculate the factor of change between the densities of Al and Cu

. Calculate the factor of change between the atomic weights of Al and Cu.

. Calculate the percent composition of the brass.

Conclusion: (essay in MLA format)

Explain completely the concepts of density, how we make assumptions about concentration, and how to approximate the atomic volume.

Example Calculations:

Aluminum:

Mass: 17.35 grams   **rounded to 3 significant digits = <u>17.4 grams</u>**

Diameter: 1.27 cm

Radius: 0.635 cm

Height: 5.07 cm

1 cubic centimeter can be symbolized as a cc or cm³

Volume = π * (r)² * H

Vol. = 3.14(0.635cm)² * 5.07cm

Vol. = 6.419 cm³   rounded to 3 significant digits = 6.42 cm³

Density = mass / volume

Dn = 17.4 gm / 6.42 cm³

Dn = 2.71 g/cm³

Percent Error = (abs difference / expected) * 100%

Error = (2.71 – 2.70) / 2.70 * 100%

Error = (0.01/2.70) * 100%

Error = 0.37 %

REPEAT ALL STEPS ABOVE FOR REMAINING SAMPLES

Number of moles in 1 cm³ = SM / MM

Al	Fe	Cu
2.7 / 27 = **0.10 moles**	7.87 / 55.85 = **0.141 moles**	8.96 / 63.55 = **0.141 moles**

Density Factor of Change:

Change = Dn of Cu / Dn of Al

Change = (8.96 g/cc) / (2.7 g/cc)

Change = 3.31 X   density increased by 331%

Molar Mass factor of Change:

Change = MM of Cu / MM of Al

James Signorelli

Change = 63.55 MM / 27 MM

Change = 2.35X   molar mass increased by 235%

Brass:

If Brass is **50% copper** and **50% zinc**, it would have an <u>average density</u> of **8.05 g/cm³**

** Since the Brass sample had a density of 8.45 g/cm³, it clearly is more than 50% copper.**

Cu	+	Zn	=	Brass
8.96 (x)	+	**7.14 (1-x)**	=	**8.45**
8.96(x)	+	**7.14 – (7.14x)**	=	**8.45**
8.96x – 7.14x			=	**8.45 – 7.14**
		1.82 x	=	**1.31**
		x	=	**1.31 / 1.82**
		x	=	**0.72 (decimal equiv of 72%)**

Analysis of Brass sample:

Brass Sample: 72 % copper   28% Zinc

Other common alloys:

       Bronze: Copper & Tin mixture

Pure gold is 24 karat:

       18 K Gold: 75% gold & 25% nickel

       10 K Gold: 42% gold & 58% nickel

In this first lab of chemistry, we study one of the discoveries and inventions of Archimedes of Syracuse. He lived from 287 to 212 BCE. His investigations included the proof that (Pi) was a number between 22/7 and 221/71, therefore, it is an unending value. He devised the formula for the volume of a sphere, he invented the "Archimedes Screw" for bringing water from one level to another, and he wrote the formula for we call Density. Density is defined as a relationship between the mass of an object and its volume. In simplicity, "Density is the mass of an **imaginary cube** having a total volume of **1.00 cm³**" (1 cm X 1 cm X 1 cm).

Archimedes has a legend associated with the discovery of density. The legend, whether real or embellished, states that Archimedes determined the purity of a gold crown by comparing the mass and volume of a bag of gold coins to the finished crown. If the mass and volume of the crown matched the mass and volume of the coins used, then the crown was pure gold. Archimedes went on to prove that no two substances <u>of his era</u> could share the same density unless they were made of the same material. Modern technology has greatly improved upon his method, but his idea is still rock solid.

This physical property of matter (density & specific gravity)) has survived the test of time. Even after 2300 years, we still use it as the first test of identity for an unknown substance. In this lab we collected data on several "unknown" materials. By using the concepts of density, we can determine the identity of each of the samples tested. Beyond that, however, we can also derive additional information, not known to Archimedes 23 centuries ago. We can use this data to deduce that atoms are not the same size. We can also use this data to count the number of atoms in that **1.00 cm³ sample**.

In the lab this week, four metals were tested. The results were used to determine the identity and purity of the three elemental samples. The data was also used to determine the percent composition of the brass alloy. An optional fifth sample can be available at the instructors discretion, if time permits. The data collected from the mass and volume determination was processed using Archimedes formula to find the derived term we call density. This value was then compared to the accepted standard found on the periodic table. The difference between our data and the accepted standard is then used to determine the percent purity of the same, as it is a direct function of the percent error. The brass sample, being an alloy, has many different possible combinations. If the brass was made of 50% copper and 50% zinc, then it would have had a density of 8.05 grams/cm³. If the alloy contained more than 50% copper, the heavier metal, then its density would rightfully be greater than 8.05. Since our sample density was determined to be 8.45, it is safe to conclude that the brass sample was much more than 50% copper.

In determining the number of atoms in the sample, we must use the standard volume of the cube being tested. The density value is the mass of a cube which is 1 cm by 1 cm by 1 cm. The number of atoms filling that cube was determined by using the formula of Amedeo Avogadro. This equation was devised in 1856 and states that the number of atoms (moles) in a sample is the mass of the sample divided by the molar mass of that material. Since our density samples were all solids, it can be assumed that the spaces between the atoms were at a minimum. This allows us to assume that most of the volume of the sample is therefore the volume actually taken up by the atoms of that sample material, and not the spaces between those atoms. The mass of this imaginary cube is the density. **The density (grams/cm³) divided by the molar mass (grams/ mole) leaves us with moles of atoms per cubic centimeter.**

We see in our data that the number of moles of atoms were higher in the copper sample than in the aluminum sample. Since both samples had the same volume, then the logical explanation must be that the more abundant copper atoms are smaller than the aluminum atoms. This allows us to "pack" more atoms into the same space. When one compares the atomic volumes listed on the periodic table to our sample data, we see that there is a confirmation of this assumption. The table confirms that copper atoms do have a smaller atomic radius than aluminum atoms. Further, since the number of **copper atoms** in one cubic centimeter is the same as the number of **iron atoms**, *<u>we can assume these atoms to be nearly the same diameter.</u>*

Another way to come to this conclusion is to look at another piece of data. The molar mass of the aluminum compared to that of the copper. The mass of copper is only 2.3 times heavier than aluminum. The density of copper is more than 3.3 times greater than that of the aluminum. If mass alone was responsible for density, then the copper sample should only be 2.3 times denser than the aluminum. However, if the atomic volume of the copper was less than that of the aluminum, then the density difference would be greater than the atomic mass difference, which it is! This leads us to a conclusion that **as the atomic mass increases, the atomic volume decreases**. This general trend is supported by the data across the periodic table.

In closing, we can see that more information has been derived from the work of Archimedes than he could have imagined 2300 years ago. Not only do we have a valid tool to identify a material, but using this with the works of other scientists, we also have insight into the size and quantity of atoms making up the materials around us. This lab points out that to an open mind, there is much more to be seen than what appears to be on the surface.

LAB #2

Investigating Avogadro's First Law (N= SM/MM)

Purpose:

Amedeo Avogadro, an Italian Mathematician, determined that 602 Billion-Trillion atoms of a material would have the same mass as the value called the "Atomic Weight" from the periodic table. This lab will investigate that concept.

Diagram:

Draw or attach a photo of the equipment in set-up mode

Procedure:

. Obtain four different types of beans: Lima, Roman, Black-Eyed Peas, and Lentils

. Fill four 50 ml beakers with each of the four bean types. Make sure to top off each beaker.

. Weigh 15 beans and set aside (Set 1)*

. Weigh 15 more beans, and set aside. (Set 2)*

. Weigh 15 more beans and set aside (Set 3)*

. Add the total mass of the three sets and then divide by 45 or 90 to determine the average mass of one bean.   ***Lentils are to have three sets of 30 beans per set**

. Weigh all of the beans from the beaker, including the 45 or 90 already weighed.

. To predict the total number of beans in the 50 ml sample, divide the total mass of all of the beans by the average mass calculated in step 6

. Manually count the total number of beans in the 50 ml sample.

. Compare your results to the prediction. Calculate the percent error.

. Repeat steps 3-10 for the remaining three beans.

DATA:

	Lima	Roman	Black-Eyed peas	Lentils
Mass - set1 (15)				
Mass - set2 (15)				
Mass - set 3 (15)				
Mass of three sets (45)				
AVERAGE BEAN mass				
Total mass of all beans				
predicted count				
actual count				
Percent error				

Calculations:

Show all math work in HSPA format.

Conclusion:

Write a brief biography of the Italian Mathematician, Amedeo Avogadro. Then, explain his work on the Mole concept. How is this lab, and the Density lab, related to his work? Explain why we use three sets of beans for our weighing analysis. How is this rooted in Probability & Statistics? The main concept of Avogadro's work is that you can count by weighing, if you know the fundamental mass to divide by.

Calculations:

Lima Beans

Sample #1 ...	16.155 g.	(15 beans)
Sample #2 ... 15.495 g.	(15 beans)	
Sample #3 ... 16.875 g.	(15 beans)	

Total mass of three samples = **48.525 g** (S1+S2+S3)

Average mass of one bean = total mass / total beans

Ave mass = 48.525 g / **45**

Ave mass = **1.078 g / bean**

Mass of all beans in 50 ml beaker = 54.98 g. (includes the 45 already counted)

Predicted count = total mass / average mass (N = SM / MM)

Predicted count = 54.98 g / 1.078 g/bean

Predicted Count = **51 beans**

Actual Count = 52 beans

Percent Error = (Absolute difference / actual count) * 100%

Error = (52 – 51) / 52 * 100%

Error = **1.92 %**

Repeat these calculations for the remaining bean samples.

Avogadro's 1st law:
N = SM / MM
Moles = sample mass / molar mass

Amedeo Avogadro was an Italian mathematician in the middle 1800's. He devised a way of measuring the concentration of particles in a sample. Whether the sample is composed of atoms or molecules, he determined that the number of particles having a mass number matching the **"atomic weight"** was always **6.02×10^{23}**. In other words, if you had a sample of carbon with a mass of 12.011 grams, then you would have an Avogadro number of atoms. If you had a sample of water with a mass of 18.00 grams, then you would have his number of molecules in that water sample. Avogadro gave us the ability to count particles (concentration) by simply weighing the sample. This made balancing equations and the application of chemistry a predictable science. Avogadro's number has become the standard in chemistry. When we look at the periodic table, and read the "mass", we are really getting the mass of six hundred and two, billion-trillion atoms.

In this lab, we take large lima beans and determine their average mass per bean. As no two lima beans are the exact same size, then they can't be expected to have the exact same mass. To negate these differences in the beans of our sample, we randomly gather and weigh three groups of beans. Each groups mass is added together. The simple math of dividing the total mass of all of the beans in the three groups by the number of beans which were weighed produces the average mass of one bean. This would be the equivalent of what is now called the **molar mass**. Now one needs only to weigh a sample of that bean and then divide by this determined average mass, to predict the total number of beans in the sample being investigated. This procedure is repeated for the other three types of beans and a pattern emerges.

A fifty milliliter beaker was used to create a fixed sample volume. In this experiment, we are trying to determine the number of beans which fit into a 50 ml sample volume. So now, for each of the four bean types tested, a fifty milliliter sample was weighed, and then the number of beans in that sample was predicted. This number is based on the average mass of that bean type. As the data shows, the method is very accurate. As the beans become smaller and lighter, they also have become more numerous. A greater number of smaller beans fits into the fifty milliliter beaker than larger beans. If we were to extend this concept down to the size of an atom, then one can see how particle counts in the trillions could be expected. Avogadro gave us the ability to draw relationships between particle size, particle mass, and particle concentration, contained in a fixed volume of a sample.

To achieve an accurate test of these relationships, a fixed volume for atoms and molecules had to be devised. If we compare the density of a sample, which is the mass of a fixed volume (one cubic centimeter) to the atomic weight (molar mass), then we have our link between Archimedes and Avogadro. How many atoms of aluminum fit into one cubic centimeter of volume? How many atoms of copper fit into that same one cubic centimeter of volume? The math shows that more atoms of copper fit into this common volume associated with density. If copper atoms are smaller, then more will fit into the 1 cc space. Looking at the periodic table, and comparing atomic radii, we see that copper atoms are definitely smaller than aluminum atoms.

In this simple lab exercise with four different bean types, we have linked atomic weight to molar mass. We have related molar mass to particle size and also particle concentration. Thus we have found a relationship between Archimedes and Avogadro across 2200 years of science.

LAB #3

Paper Chromatography of Ink

Purpose:

To separate an ink sample into its components by paper chromatography

Diagram:

Draw or attach a photo of the equipment in set-up mode

Procedure:

- Prepare a **polar solvent** by mixing equal parts of Methanol, 0.1 M. Acetic Acid, and distilled water.

- Place a drop of the sample onto each paper strip using a capillary tube.

- Place the solvents into the tube to the required depth of 1 inch. Label the tubes for the type of solvent in use.

- Insert the paper strip with the ink dot and **replace the cap.**

- Allow the action to continue until the solvent **almost reaches the top** of the strip..

- Remove each strip from the tube, dry, and measure the **Rf** for each component.

Data:

Polar Solvent

Distance solvent traveled _____ mm

Distance components traveled:

A = _____ mm (lightest)

B = _____ mm

C = _____ mm

D = _____ mm

E = _____ mm (heaviest)

Rf of A = _____

Rf of B = _____

Rf of C = _____

Rf of D = _____

Rf of E = _____

Rf = distance sample travels / distance solvent travels

Conclusion:

the following are points to be addressed as paragraphs, nit fill in answers.

.   Explain the principles behind paper chromatography. Refer to the Surface tension formula from the board.

   (hint: capillary action, adhesion, cohesion)

.   What explains why the color band is more like a large blot rater than a narrow band?

. Why does a polar solvent have the greatest effect on the ink? What does this indicate about the sample?

. Why are solvent mixtures used rather than one solvent?

. Why must this procedure by done in a sealed system?

James Signorelli

Paper Chromatography

The process of separating one molecule from another in a complex sample often requires chromatography. The analytical method relies on the mass and size differences of the materials in the mixture. Physics shows us that the mass is inversely proportional to the velocity, at a constant temperature (KE). Since heavier molecules will move slower than lighter ones, in a given period of time, we get separation.

The first requirement for this method to work is that a solvent can migrate up the paper. This happens because one material (the solvent) adheres to the fibers of the paper. These fibers are very close together. The solvent pulls itself up the fibers by capillary action. If the solvent and solute molecules are both compatible (polar or non-polar), then something else happens. The solute dissolves into a similar solvent. As the solvent goes up the paper, it carries the dissolved solute molecules with it. The lighter mass molecules travel faster and therefore farther in a given period of time. The heavier molecules are slower and therefore don't travel as far. If a certain material is either too heavy or too large to fit between the fibers, then it doesn't move at all.

When the solvent is migrating up the paper, it tends to evaporate off of the paper, as more surface area is "wet down" by the solvent. To prevent this evaporation, the entire process must be conducted in a sealed cabinet (test tube). The atmosphere in the tube becomes saturated with solvent vapor, making it impossible for the solvent to evaporate off the paper. The solvent continues up the paper, transporting rather than stranding, any of the solute molecules with it.

In time, the lighter molecules are found at the top of the paper, while the heavier molecules are near the middle or bottom of the paper. The solute molecules, cohering to their similar molecules forms bands on the paper. These bands identify the location of the various materials which had comprised the mixture which was analyzed.

The last act is to measure the distance the solvent had traveled, and compare it to the distance traveled by each component. This creates the Rf for each component. The Rf is a value unique to that component, in that specific solvent. It can often be used to identify the component. It is a number, and, like a fingerprint, it is unique to that material.

In conclusion, chromatography is a method of isolation and purification. Complex mixtures can be separated by their mass, size and boiling points. What is required is a solvent which is compatible with the solute. The interplay of each must also take into consideration the polarity of the materials. Next, a medium for this separation must be chosen. Chromatography can be used with mediums such as paper, coated glass plates, high pressure liquids, and finally a gas carrier agent passing through a heated bed of neutral material. The last part of the method is time. The method requires time for the separation to occur. If given enough time and distance, one could even separate isotopes of the same element.

LAB #4

Inverse Square Law

Purpose:

To determine the amount of radiation detected from an Alpha source at varying distances and to compare these values to predictions using the Inverse Square Law.

Diagram:

Draw or attach a photo of the equipment in set-up mode

Procedure:

.   Determine the back ground count for 5 minutes.

.   Set the source 8 cm from the geiger counter and record the highest reading.

.   Move the source to 16 cm, and repeat step #2.

.   Move the source to 24 cm and repeat step #2.

.   Move the source to 32 cm and repeat step #2

.   Move the source to 40 cm and repeat step #2

.   Move the source to 48 cm and repeat step #2

.   Move the source to 56 cm and repeat step #2

.   Move the Source to 64 cm and repeat step #2.

. Calculate the expected **CPM** at each distance using the Inverse Square Law and the count from step #2.

. Graph the data using the relative distance for the X-axis and the counts per minute for the Y-axis.

Distance	Rel. Distance	Expected CPM	Actual CPM
8	1		
16	2		
24	3		
32	4		
40	5		
48	6		
56	7		
64	8		

Calculations:

Show the data for the expected column.

Graph the data using the Relative Distance as the "X" axis and the Actual & Expected counts for the "Y" axis. Refer to the graph provided as a reference.

Conclusion:

Make a drawing of the principles of the Inverse Square Law and explain the drawing.

Explain how energy spreads as it goes away from the source. Make a list of other forms of energy which follow the Inverse Square Law. Explain each example.

All radiant energy spreads away from a point source at an even rate. The increase may be referred to as the diffusion of energy. We are taught that diffusion is the spread of something from an area of great concentration, to least concentration. Does this concept also apply to energy? The answer is yes! Energy and matter spread out from an area of great concentration to lesser concentrated areas. Stars give off light because they are the great concentration of that form of radiant energy. There are no stars that emit dark! When it comes to all forms of electromagnetic radiation, they all obey the same laws of physics. One of these laws is called the Inverse Square law.

As energy spreads away from a source, it covers a greater and greater area. Think of a flashlight held against a wall. As the light is pulled away from the wall, the beam grows wider. Draw a circle on the wall, and hold the flashlight near the circle. Pull the light away from the circle, the beam of light spills out from that circle. You are illuminating more and more of the wall. How much light is actually hitting the circle now? The circle is your detector. Since less light is in the circle at greater distance, the circle is darker. It records a decrease in the energy striking the detector (circle).

Photography uses this principle. As the distance to the subject increases, less and less light enters the camera lens. The image gets darker. To correct for this, the camera opens the lens wider (f stop) and / or keeps the lens open longer to correct for the decreasing light. As you drive West on route 80, a New York radio station is more difficult to hear. The energy from the radio transmitter is spreading and the car radio loses the signal in time. In fact, the signal is still there, but, so weak that the radio can't amplify it enough to he heard. The signal is lost in the background static.

In this lab, a radioactive isotope is placed eight centimeters from the Geiger counter. The Geiger counter detects the energy streaming from the isotope. The detector can't get bigger. As a result, as the isotope is placed farther away from the Geiger counter, it records less strikes per minute on the detector. Each distance used in the lab must be a multiple of the original eight centimeters. Does the energy change agree with the math of the Inverse Square law? If yes, then the law should allow us to predict the level of energy at each distance from the detector and the isotope. If the law does not apply, then I will see no predictable data. We then graph the predicted data against the actual data to see a relationship, free of minor fluctuations in the isotope. If the two graphs overlay each other, then we have confirmation that the law applies in this experiment. If the two graphs are completely different, then we have confirmed that the law did not apply in this case.

Since our graphs virtually lay on top of each other, we have proof that the Inverse Square Law applied. Our Geiger counter proved that the energy streaming from the Isotope did in fact spread in three dimensions. The **area covered** by that energy grew **GEOMETRICALLY** as the distance to the detector increased. It was proven that radioactivity obeys the inverse square law, just as all energy must obey. It is tied to the laws of Diffusion. Matter and energy obey the same laws. Einstein was correct, Matter and Energy are related concepts.

12,800	12,800
3200	3150
1422	1400
800	775
512	500
356	325
261	250
200	190

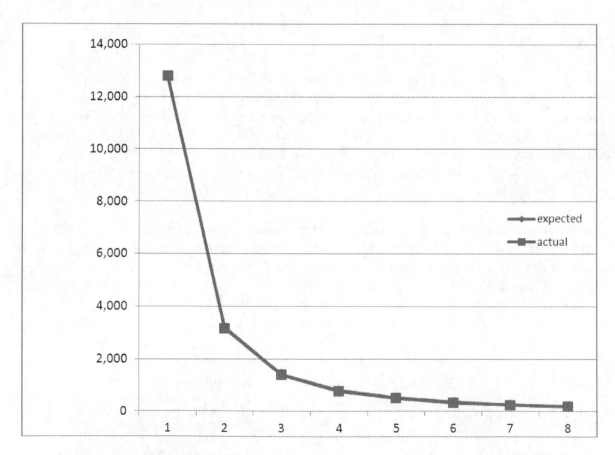

LAB #5

Determining Spectral Lines for Hydrogen

Purpose:

using the formulas of James Balmer, Max Planck, and Albert Einstein, one can predict the bright spectral lines for any element. In this exercise, we will determine the four <u>visible lines</u> for Hydrogen gas.

Diagram:

Draw or attach a photo of the equipment in set-up mode

Procedure:

. Using the three equations supplied in class, and the ionization energy for hydrogen, calculate the photon energy, the frequency, and the wavelength of each spectral line produced.

. Complete the table with your answers to the calculations.

. On a clean piece of graph paper, in landscape mode, draw an 8 inch line. Mark every two inches accordingly (4, 5, 6, 7, 8)

. At each one inch line, place an unlabled short line.

. At each ½ inch line, place a smaller unlabled shorter line than used in step #4.

. Label this chart **wavelength in X 10^{-5} cm**

. Draw a line at each proper location on your chart that matched the wavelength value you calculated in the chart in step2 1 & 2.

- Set up the spectral discharge apparatus.

- Install the Hydrogen tube and turn unit on

- Place a spectroscope in front of the glowing hydrogen tube.

- Observe the bright-line spectral pattern. Record the locations of each line on the grid in the scope.

DATA:

complete the table

Calculations:

Show all math used to complete the table, using standard math format.

I.E. for Hydrogen is 313.6 kcal/mole all other elements may be found on the periodic table

$En = I.E. [1/(2)^2 - 1/(N)^2]$	$N = 3, 4, 5, 6, 7, 8$	
$v = En /h$	$h = 9.54 \times 10^{-14}$	Planck's constant
$\lambda = C / v$	$C = 3.0 \times 10^{10}$ cm/sec	
$\lambda = h/ M*V$	M = mass of electron	V = velocity of electron

Conclusion:

Explain in detail the basic concepts of Quantum Mechanics.

Balmer: $N = 3$ *(repeat with 4, 5, 6, 7, 8)*

$En = I.E (1/2^2 - 1/N^2)$ (IE = 313.6 kcal/mole)

$En = 313.5$ kcal $(1/2^2 - 1/3^2)$

$En = 313.5$ kacl $(1/4 - 1/9)$ find a common denominator

$En = 313.5$ kacl $(9/36 - 4/36)$

$En = 313.5$ kcal $(5/36)$

$En = 43.6$ kcal   (photon energy)

Planck:

$$En = v * h \qquad\qquad (v = \text{frequency of light}) \qquad (h = 9.54 \times 10^{-34})$$

$$v = En / h$$

$$v = 43.6 \text{ kcal} / 9.54 \times 10^{-14}$$

$$v = 4.56 \times 10^{14} \text{ cycles per second}$$

Einstein:

$$C = \lambda * v \qquad\qquad (C = 3.0 \times 10^{10} \text{ cm/sec}) \qquad (\lambda = \text{wavelength})$$

$$\lambda = C/v$$

$$\lambda = 3.0 \times 10^{10} \text{ cm/sec} / 4.56 \times 10^{14} \text{ cycles/sec}$$

$$\lambda = 6.58 \times 10^{-5} \text{ cm/cycle}$$

$$\lambda = h / M*V \qquad\qquad C = v*\lambda \qquad\qquad En = h*v \qquad h = 6.68 \times 10^{-31} \text{ m/s/mole}$$

$$V = h/ M*\lambda$$

$$V = 6.68 \times 10^{-31} / 9.11 \times 10^{-31} * 6.58 \times 10^{-8}$$

$$V = 111,438 \text{ m/s}$$

Repeat all calculations for N = 4, 5, 6, 7, 8

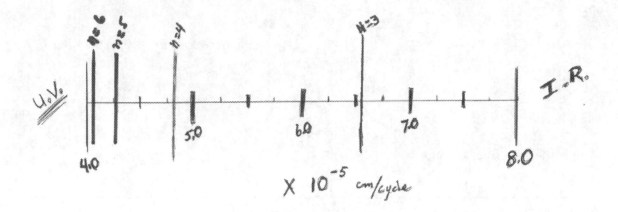

N	En	Frequency	Wavelength	color
	Kcal/mole	cycles/sec	cm/cycle	
3	43.6	4.57 E 14	6.56 E -5	Red
4	58.8	6.18 E 14	4.86 E -5	Cyan
5	65.9	6.92 E 14	4.34 E -5	Blue
6	69.7	7.32 E 14	4.10 E -5	Violet
7	72.0	7.56 E 14	3.97 E -5	UV
8	73.5	7.70 E 14	3.89 E -5	UV

The study of electrons in orbit around their nucleus is called **Quantum Mechanics**. In this unit, the energy required to move an electron from one level to another is investigated. Then, the frequency and wavelength of the electromagnetic radiation emitted is calculated. The velocity of the electron is also determined. Electrons act like any other particle of matter in an orbit around a center of attraction. In physics, one studies planets in orbit around a gravitational mass. In atoms, the "satellite" is an electron orbiting around the nucleus, which is the center of an oppositely charged mass.

The investigation begins with Johann Balmer in 1885. He calculated the **radiant energy** emitted by an electron dropping from a higher orbit to a lower orbit in the Hydrogen atom. Balmer's formula lead to a basic understanding in Astronomy. It proved that stars are not the same temperature, nor are they stationary in their positions. This first formula used in **quantum mechanics** proves that whether an electron is "jumping" up or "falling down" in energy levels, the transition involves the exact amount of energy. When an electron is **excited,** it increases in velocity, and "jumps" to a higher orbit. When the electron "cools off", it slows down, and returns to the original orbit. This original location is called the **ground state**. The energy given off is often perceived as visible light. The amount of energy is a certain quantity called a **"photon"**. It is the basis of **quantum energy.**

Next, the formula of Max Planck was applied to determine the frequency of light associated with a "quanta" of energy. Planck was awarded the Nobel Prize in 1918 for his work on Quantum Mechanics. His work determined that as the energy of a photon increased, it created light of a greater frequency. The transition from low energy to higher energies causes the **spectrum** of colors. Red light is produced by the lowest energies, and violet light involving the highest energies. Planck's equation is not restricted to the work of Johann Balmer and visible light. It is universal to all of quantum mechanics. Lymann worked on Ultraviolet light. Paschen worked on Infra Red light. Brackett, Pfund, and Humphries all worked on different series of light, visible and invisible, emitted by excited Hydrogen electrons. Their work spanned the light spectrum from IR, to visible, to UV, to X-Rays. Planck's equation can be used to calculate the frequency of light emitted by an electron regardless of the series.

The third equation used in this lab was that of Albert Einstein. It shows the inverse relationship between **frequency, wavelength**, and the **speed of light**. This formula demonstrates that red light has the longer wavelengths and blue light the shortest. As the waves get shorter and shorter, they enter the ultraviolet area of the electromagnetic spectrum. *"All radiant energies on the electromagnetic spectrum travel at the speed of light".*

Upon inspection of the table created by the application of these three laws, the data is demonstrating their connection. As the <u>**energy level**</u> increased, the <u>**photon energy**</u> increased. As the photon energy increased, the <u>**frequency of light**</u> increased. As the frequency increased with increasing energy, the <u>**wavelength of light**</u> was decreasing. In addition, the <u>**velocity of the electron**</u> was increasing. This follows the basic laws of physics. *The orbital radius of a body is defined by the velocity of the object.* This lab points out that all bodies in orbit must conform to the same laws of physics. It matters not whether that body is a planet orbiting a star, a communication satellite orbiting a planet, or an electron orbiting a nucleus. *All bodies in orbit adhere to the same laws of physics!*

The final observation from this lab has to deal with the **spacing of the energy levels**. It is observed that the spectral lines are not equally spaced. The spectral lines in the blue region are closer than those in the red region. As the energy is increased, the electron velocity does increase. However, notice the amount of each increase! The outer orbits are not as far apart from each other as the inner orbits. If one extends this concept, it is expected that a finite number of elements can be created. This is because the super heavy elements would have outer electron levels too close to a previous level to orbit in a stable path. Presently, science can create new elements. Sometime in the future, this will not be possible. We will approach and attain an atomic size where the orbital spacing prevents the formation of new elements.

Balmer Series
H2 Spectral Lines

N	En	Frequency	Wavelength	electron velocity	color
	Kcal/mole	cycles/sec	cm/cycle	meters / second	
3	43.6	4.57×10^{14}	6.56×10^{-5}	11,177,746	**Red**
4	58.8	6.18×10^{14}	4.86×10^{-5}	15,087,657	**Cyan**
5	65.9	6.92×10^{14}	4.34×10^{-5}	16,895,395	**Blue**
6	69.7	7.32×10^{14}	4.10×10^{-5}	17,884,393	**Violet**
7	72.0	7.56×10^{14}	3.97×10^{-5}	18,470,029	UV
8	73.5	7.70×10^{14}	3.89×10^{-5}	18,849,875	UV
ground state N=2					
IE of H_2 = 313.6 Kcal/mole					
C = 300,000,000 m/s					
IE = 313.6 kcal/mole					

LAB #6

Decomposition of Sodium Bicarbonate

Purpose:

All chemical reactions can be "mapped" using the mass of the reactants to predict the mass of the products. This is called MASS BALANCE. In this experiment, 2.00 grams of the reactant Sodium Bicarbonate (Baking Soda) will be heated. The products of the decomposition will be Sodium Hydroxide and the gas, Carbon Dioxide.

Diagram:

Draw or attach a photo of the equipment in set-up mode

Procedure:

. Weigh a clean, dry ceramic cup called a **Crucible.** Record the mass as M1

. **Tare the balance (re-zero).**

. Add 2.00 grams of Sodium Bicarbonate to the cup

. Place cup onto the ring stand in the wire triangle

. Light your bunsen burner and adjust the flame to be the nearly invisible blue double cone.

. Heat the crucible for five minutes. Remove the burner **BUT DO NOT TURN IT OFF!!**

. Allow the crucible to cool before reweighing it. (You could melt the balance as the cup is over 900 degrees)

. Calculate the mass of the chemical in the cup.

. Return to the ring stand and heat an additional five minutes.

. Turn off the burner.

. Allow the sample to cool, and reweigh.

. Calculate the mass of the sample in the cup.

. Compare to the expected mass.

Data:

Mass of dry, empty crucible	_____ grams	
Mass of sample	__2.00____ grams	
Mass of crucible + sample	_____ grams	
Mass of crucible after 1st heating	_____ grams	
Mass of crucible after 2nd heating	_____ grams	
Mass of sample in crucible	__0.973____ grams	(mass of NaOH remaining)
Mass lost in crucible	_____ grams	(mass of CO_2 discharged)
Expected mass of product	__0.952 __ grams	
% Error	_____ %	

Calculations:

Show the reaction

Show the calculations for the molar mass of each material in the reaction

Show the math for the calculation of the expected mass

Show the calculation for the percent error

Conclusion:

This is a decomposition reaction. Explain the mass balance of this reaction.

Mass Balance of Reaction

NaHCO$_3$	$\longrightarrow$	**NaOH**	+	**CO$_2$**
1 mole		**1 mole**		**1 mole**
84 grams		40 grams		44 grams
SM		**expected mass**		expected loss

Molar Mass of Reactant & Products

NaHCO$_3$ = 84 g/mole   **NaOH = 40 g/mole**   **CO$_2$ = 44 g/mole**

$23+1+12+ (16*3) = 84$ $23+16+1 = 40$ $12+ (16*2) = 44$

Proportional Mass Balance

$$\frac{\text{NaHCO}_3}{\text{NaOH}} = \frac{84 \text{ gms}}{40 \text{ gms}} = \frac{2 \text{ gms}}{X \text{ gms}}$$

expected mass
x = 0.952 gms. (ash)

$$\frac{\text{NaHCO}_3}{\text{CO}_2} = \frac{84 \text{ gms.}}{44 \text{ gms.}} = \frac{2 \text{ gms.}}{y \text{ gms.}}$$

expected mass
y = 1.047 gms. (lost)

Error Percent:

Error = (absolute difference / expected) * 100%

Error = [(0.973 − 0.952) / 0.952] * 100%

Error = 2.21 %

Thermodynamics		ΔH	ΔS
$NaHCO_3 \rightarrow$	$Na + \frac{1}{2} H_2 + C + 1\frac{1}{2} O_2$	+950.9	+ 0.102
$Na + 1/2 O_2 + \frac{1}{2} H_2 \rightarrow$	**NaOH**	-469.0	-0.646
$C + O_2 \rightarrow C\,O_2$		-393.5	+0.2137
$NaHCO_3 \rightarrow$	$NaOH + CO_2$	+88.25	+0.2511

Reversing Temp:

When ΔH & ΔS have the same sign, then the reaction may reverse

T = ΔH / ΔS

T = +88.25 / +0.2511

T = 351.5 Kelvin

This reaction reverses when heated to 351.5 K or (78 C.)

$\Delta G = \Delta H - T\Delta S$

$\Delta G = (+88.25) - (298 * 0.2511)$

$\Delta G = + 9.42$ kj (non-spontaneous)

Summary: **endothermic, increasing entropy, <u>non-spontaneous</u>, & reversible**

The decomposition of **Sodium Bicarbonate** ($NaHCO_3$) upon heating is applied on a daily basis in the baking and frying of certain foods. The products of the decomposition are **sodium hydroxide (NaOH)**, a very bitter caustic material, and **carbon dioxide (CO_2)**, a gas. The CO_2 gas is discharged by the reactant as it is heated. These gas bubbles diffuse through the food causing it to rise. These CO_2 bubbles are responsible for the holes we find in bread, pancakes, and muffins. The sodium hydroxide remaining in the baked product imparts a bitter taste. To compensate for this, sugar is often added to the recipe.

The chemical reaction taking place is **endothermic** because energy must be added for this decomposition to occur. The reaction cannot progress unless heat energy is constantly added. It can't continue on its own if one stops heating it, therefore, it is **non-spontaneous**. Reactions which create more randomness (increased **Entropy**) are favored in nature. However, nature also prefers exothermic reactions. An exothermic reaction produces heat at is progresses, is self sufficient, and spontaneous. Because this decomposition reaction is endothermic and the randomness increased, it is **reversible.** A similar reaction which is both endothermic, increases in entropy, and therefore reversible, is a process we call **"Photosynthesis".**

In the late 18th century, a French chemist, **Antoine Lavoisier**, discovered the **law of conservation of mass**. He wrote that in a closed system, the products will weigh exactly what the reactants weigh. This lab did not have a closed system. The gas was allowed to escape to the atmosphere. However, the mass of the remaining NaOH is proportional to the beginning mass of $NaHCO_3$. In this lab, a sample of sodium bicarbonate was heated. The gaseous carbon dioxide was driven off leaving the sodium hydroxide (ash) in the crucible. All matter has mass. The mass lost by heating the contents of the crucible represents the mass of the carbon dioxide driven off. The mass of the sodium hydroxide remaining in the cup is the balance of the sample mass.

John Dalton wrote in the early 1800's that "matter joins in a **definite composition**". This law allows us to calculate the percent composition of all molecular substances. By applying the law of definite composition, we determine that the sodium bicarbonate is slightly more than **52%** carbon dioxide. Therefore, it can be assumed that the sample, upon heating, will lose 52 % of its mass. The ash remaining will be slightly less than **48%**. Adding these two values yields **100%.** Thus, Lavoisier's law was correct and all mass is conserved and can all be accounted for. John Dalton was also correct. The definite proportion of the reagent allows us to predict both the mass lost and the mass remaining.

The error in this experiment was slightly over two percent. The **two major causes for error** in an experiment of this design are found in the **precision of the balance** used to calculate the mass of both the reactant and the remaining product. It is always preferred to use a digital analytical balance, when possible. The other source of error is often in the heating of the material. If the mass lost is less than expected, then more heating is required. The extra mass implies that not all of the carbon dioxide has been driven out of the sample. If the mass remaining is less than the mass expected, then the cause could be the **source of the heat**. If using a Bunsen burner, which functions at 1200 + degrees Celsius, then one can expect to drive off some of the ash residue in the intense flame and its updrafts. ***To reduce this loss, a cover on the crucible is required.***

In closing we can see that a set of laws written two hundred years ago, using almost no technology, can still be proven today. Matter has very definite properties, including conservation of mass and the materials are reacting in definite proportions. In the modern era, we can easily prove these laws with our technology. We now assume these laws to be in effect and we build upon them when planning future reactions.

LAB #7

Formula of a Hydrate

Diagram of Equipment:

ring stand, triangle, crucible, Bunsen burner OR **HOTPLATE**

Purpose:

To investigate the hydroscopic property of certain salts, called a **HYDRATE.** This is a Decomposition Reaction. The addition of heat will drive off the water leaving an **ANHYDRATE.**

Procedure:

- Weight a clean, dry crucible to <u>several decimal places.</u>

- <u>Zero the balance.</u>

- Add **2.50 grams** of **copper sulfate pentahydrate** (blue crystals) to the crucible.

- Heat for several minutes until the blue color disappears, and a white powder (ash) remains. *Use the glass rod to stir the powder to insure uniform dryness.*

- Transfer to the drying chamber *(dessicator)* for several minutes for cooling.

- Reweigh crucible, subtract the orig. crucible mass, and determine the mass of the white powder. *This mass is proportional to the mass of the Hydrate.*

- Using a supplied eyedropper, slowly add water, drop by drop and observe what happens. *You should see the steam coming off of the powder.*

- Feel the bottom of the crucible. *It should be quite hot!*

Data:

Mass of dry crucible:	_____ grams
Mass of blue crystals:	_2.500_ grams (SM)
Mass of crucible + crystals:	_____ grams
Mass of dried ash:	_____ grams
Expected mass of ash:	_1.600_ grams (expected mass)
Percentage error	_____%

Calculations:

* Show the entire mass - balanced Reaction

* Show all work in HSPT/HSPA format to complete the data table

Show the complete mass-balance reaction map

Conclusion:

Molecules that attract and attach to water express a **hydroscopic** property. These molecules become hydrated, naturally. The **hydrate**d crystal is heated to drive off the water. The remaining residue is called the **anhydrate**. The reaction to drive off the water is **Endothermic** and **Non-spontaneous**. The reaction is reversible. The reverse reaction, in attaching the water is **Exothermic** and **Spontaneous.**

All spontaneous reactions are exothermic and they are favored in nature.

Explain all of the above major points.

Hydrate Lab

LAB #7

Formula of a Hydrate:

Many molecules exhibit a property whereby they can absorb water. They can extract the water vapor from the air and attach it to the surface of the compound. This property is called **Hydroscopic.** All acids, bases, and salts have this property. When they come into contact with water, in any form, they will bond to the water. If these compounds remove the water from skin, they create a chemical "burn." Many of these compounds simply remove the water from the air. Any compound which has already come in contact with water, and it is now attached to the surface, is called a **Hydrate.** The amount of water that may attach to the molecule is a property of that molecule. In this lab, copper sulfate pentahydrate was heated to drive off the five moles of water adhering to the molecule. The dried compound, free of water, is called the **Anhydrate**. This literally means "without water."

Raw rice is often placed into salt shakers in restaurants because the rice is more hydroscopic than the salt. The rice prevents the salt from becoming a solid mass and therefore impossible to flow out of the shaker. Vitamins also have a salt pack placed into the bottle to protest the pills from moisture, when the bottle is opened.

Reactions are classified by their **Thermodynamics.** The change in the energy content of a reaction is called its Thermodynamics. It means the change in heat. All naturally occurring reactions are **spontaneous**. Nature prefers them over all other reactions. All spontaneous reactions are also exothermic. **Exothermic** means to give off heat (energy). *All exothermic reactions are spontaneous and preferred in nature.* Reactions that absorb heat so they can react are called **Endothermic** reactions. These reactions are also **non-spontaneous**. You must apply energy in some form of they can not take place. An example of an endothermic reaction is bread baking. If the oven is not turned on, if the dough is not heated, you don't get bread.

When an anhydrate picks up water, it is a spontaneous process. It is therefore also Exothermic. A Hydrate gains water in an *exothermic, spontaneous reaction*, which is favored in nature. When we drive the water off, it is an Endothermic, non-spontaneous process. **The loss of water by a hydrate is not favored in nature**. It cannot take place on its own.

If a reaction process can be reversed, then everything done to the system initially is reversed. If heat is added to drive off water, then, when water is added back, the heat is driven off. This was proven in the lab when after the sample had been dried, and some water was added back. The blue color of the hydrate returned. The sample got very hot, as it now had to shed the energy it picked up during the drying process. The fine white powder of the anhydrate was converted back into the blue solid of the hydrate. **Everything in the first reaction had reversed!**

LAB #8

Lead Iodide Lab

This week, the lab exercise had us investigating a double replacement reaction between two solutions. The first solution was 0.2 molar Lead Nitrate. The other solution was 0.2 molar Sodium Iodide. The determination of how and why they reacted, and as a result, the prediction of the **Theoretical Yield** requires several logical assumptions. We begin by analyzing the Lead ions generated by the first solution, the Lead Nitrate. The element Lead has two possible **oxidation states.** A complete Mass-Balance Analysis allows us to predict the outcome of the reaction.

These states are (+2) and (+4). When you determine the **electron configuration** for the element Lead, we find that it has four valence electrons in the 6th energy level. These four electrons are found in two sub-levels. A pair of electrons are in the 6s sub-level, and there are an additional two unpaired electrons in the 6p sub-level. The two unpaired electrons are easily removed in a chemical reaction run at or near room temperature. The Lead atom would express a (+2) oxidation state. However, if the reaction were heated, then the two paired electrons in the 6s sub-level could also be driven off, giving the Lead atom a (+4) oxidation state. Since we used chilled solutions from the refrigerator, for reasons to be explained later, *it is assumed that the Lead would express the (+2) oxidation state.* Only the two unpaired electrons from the 6p sublevel would be exchanged in this reaction. The two possible outcomes for the product can be tested. The PbI_2 product, formed with chilled solutions, is yellow in color. The PbI_4 product, formed during heating the solutions, is orange in color.

The Iodine atom had several possible oxidation states. They are **(-1),** (+1), (+3), (+5), and (+7). Since the Lead atom has a positive charge, *it is logical to assume that the Iodine must have the (-1) oxidation state in this reaction.* With these two assumptions, one derives the combining ratio between Lead and Iodine to be 1:2 in this reaction. This lab will combine 50 ml of the lead solution with 100 ml of the iodine solution. The product expected (PbI_2) is slightly soluble in water above 15 Celsius. For this reason, both solutions and all water used for washing the product are kept refrigerated. An effort must be made to not exceed this 15 Celsius threshold. Further, by keeping the volumes at 150 ml in this experiment, we eliminate the total water volume as a factor in the **percentage yield** of the product. If volumes for all lab groups are equal, then the losses to solubility **(Ksp)** are under control. *These losses would be the same in each group.*

After all of the lab groups samples were made and were filtered, the product was dried overnight. Rapid heating in an oven to dry the samples was to be avoided, as this could cause oxidation of the Lead compound, thereby changing our results. All of the filters were washed sufficiently, so all of the excess Lead or Iodine was removed from the precipitate before drying. The sodium nitrate by-product was also washed free of the precipitate. This reaction relies on **Collision Theory** to produce a product. Anything which can guarantee a collision of the reactants will improve the outcome of the reacton.

Heating the reaction would improve the collision of the reactants, but it would also make the precipitate soluble. In addition, the combining ratio for heated lead iodide is greater than the colder solution ratio. In trying to improve the reaction outcome by heating, you would in fact alter the hypothetical yield.

In closing, let us summarize the concepts covered by this exercise. All atoms rely on *Collision Theory* to react. Anything that supports these collisions will increase the amount of final product. Materials that appear *insoluble* are still slightly soluble, unless certain precautions with temperature and total volumes are taken. Atoms join in specific combinations. Robert Dalton called this the *Law of Definite Proportions*. Any deviation from this exact ratio creates a *Limiting Reactant*, which limits the amount of product formed by that reaction. This ratio is determined by the *oxidation states* of the atoms reacting. *Temperature can influence* both the oxidation state expressed by an atom, and the *solubility* of the reaction product. And finally, by comparing the Mass-Balance of the reaction, we can see if a law is at work, and how much of the reaction has actually run to completion.

LAB #8

Title: **Formula of a precipitate**

Purpose:

To investigate the theoretical yield of Lead Iodide and the formula of the compound

Diagram:

Draw or attach a photo of the equipment in set-up mode

Procedure:

Day #1

.   Mix 50 ml of a **0.2 Molar solution of Lead Nitrate** with 100 ml of a **0.2 Molar Sodium Iodide** solution.

. Transfer to an ice bath for 10 minutes while you set up the filtering apparatus.

. Obtain a piece of #1 Whatman Filtering Paper and weigh it.

. Pour the mixture from step #1 onto the paper. The filtrate should be colorless and clear. The yellow lead iodide should be trapped on the paper.

. Wash several times with **chilled distilled water** to remove by-products and unreacted materials.

. Place filter paper onto a watch glass for overnight drying.

Day #2

. Reweigh filter paper (final mass) and calculate the actual yield. Compare to the theoretical yield.

Data:

Mass of Filter paper (I)   _____ grams
Mass of Filter paper (F)   _____ grams
Actual Mass of PbI_2 ppt.   _____ grams
Expected Mass of PbI_2 ppt.   _____ grams
% error   _____ %

Calculations:

* Show the entire Mass Balanced Reaction.

* Show all work in HSPT/HSPA format

Conclusion:

The *__actual mass__* of the precipitate produced is rarely as high as the *__expected mass__*.

A reason for this phenomenon is based on **"Collision Theory".**

Explain how this theory was part of you experimental results.

This experiment required the use of cold water and chilled solutions to work properly.

Why is a cold temperature important in this reaction and for the filtering process?

LAB #9

Calculating the Molar Mass of Two Gases

Purpose:

Using the Ideal Gas Law, the determination of the molar mass of several gases will be accomplished. This exercise uses the two laws of Avogadro with the Ideal Gas Law. Because the gases are weighed in air, Archimedes laws are also used to calculate the Buoyant Force acting on the gases collected.

Diagram:

Draw or attach a photo of the equipment in set-up mode

Procedure:

. Record the temperature and Pressure of the room.

. Calculate the actual density of the air in the room using these values and the air standard of 1.29 grams/liter @ STP.

. Set up the water tray collection system.

. weigh a clean, DRY, ***collection bag******

. Fill the bag with the sample gas. Replace the rubber bulb.

. Reweigh the bag to calculate the apparent mass.

. Transfer the gas to the ***collection chamber***.+++

. Record the displaced volume of the sample gas.

. Calculate the Buoyant Force using the determined volume.

. Calculate the actual mass of the gas

. Calculate the molar mass using the Ideal Gas Law

Data:

	Propane	Methane
Room Temperature	_____ Celsius _____ K (T)	_____ **Celsius**
Room Pressure	_____ torrs _____ ATM's **(P)**	_____ **Torrs**
Density of Air (today)	_____ grams /liter	_____ g/liter
Mass of bag, initial	_____ grams	_____ grams
Mass of Bag, final	_____ grams	_____ grams
Apparent mass of gas	_____ grams	_____ grams
Volume of gas	_____ liters(V)	_____ **liters (V)**
Buoyant Force	_____ grams	_____ grams
Actual mass of gas	_____ grams **(SM)**	_____ **grams (SM)**
Actual Molar Mass	_____ g/Mol**(MM)**	_____ **g/mole(MM)**
Expected MM	_ 44_ g/mole	_16_ g/mole
Percent Error	_____ %	_____ %

Calculations:

Show all work in HSPA format

Bouyant Force = Vol * d2

Act mass = Bouyant Force + Apparent mass

$$d_2 = (d_1 * P_2 * T_1) / (P_1 * T_2)$$

$$MM = (SM * R * T) / (P * V)$$

Conclusion:

Avogadro's hypothesis states that one mole of any gas occupies 22.4 liters at STP. The volume is a constant, if the number of molecule is also a constant, at the same temperature and pressure? Develop an explanation of Avogadro's 2nd Law using references to the kinetic energy of the individual particles, the collisions of the particles, the attractive forces of these particles, etc. In addition, explain why the buoyant force is important when weighing gases, but not significant when weighing solids and liquids.

Remember: it is not about the size or mass of the gas particle, it is all about the number of particles!

If it sounds like you are using <u>K-M theory</u> to explain it, you're already half way there!

*** The ***<u>collection bag</u>*** is a zip lock bag attached to a one hole rubber stopper with a automotive hose clamp. The hole in the stopper has a glass eye dropper with the rubber bulb.

+++ The ***<u>collection chamber</u>*** is a one liter graduate cylinder, filled with water, inverted in an overflow trough. As the gas is transferred into the trough nipple, the bubbles displace an equal volume of water in the cylinder.

Molar Mass of a Gas

The study of gases is important in understanding the nature of rapidly moving particles which have incredible distances between them. There are a group of properties assigned to gases called **Kinetic-Molecular Theory**. This theory states that:

- Gases consist of very small particles, each having a definite mass.

- The distance separating gas particles are relatively large, compared to the actual size of the gas particle. Basically, gases are mostly empty space.

- Gas particles are in constant, rapid, random motion.

- Collisions of gas particles with each other or the walls of their container are perfectly elastic. These particles don't lose energy or velocity from these collisions.

- The average Kinetic Energy of a gas particle depends only on the temperature of the gas.

- Gas particles exert no force on each other. Attractive forces are so low, that the sample does not slow down and condense into a liquid.

Gases are also considered to be **fluids**, in that they can flow from one place to another. The driving force of this motion is of course the kinetic energy contained by the gas. It may manifest itself as diffusion. It may also be called "wind" in larger samples, such as the atmosphere of planets. One property of all gases is that they have no definite shape or volume. They expand to fill their container in both shape and size. If that container is opened, then the gases escape to expand indefinitely.

Hundreds of years ago, a scholar named **Blaise Pascal** formulated three laws about fluids. His third law is the most profound. It states that for an object submerged in a fluid, it loses the weight of the fluid it displaces. This simply stated means that if an object is in a fluid, it pushes out the fluid so the object can occupy that space. It return, the fluid pushes back against the object, which in turn partially or completely supports that object. Therefore, it loses weight!

In this lab, a gas sample was collected in a bag and weighed. The bag is in a room full of air. This means that the bag is submerged in the air (a fluid) in the room. Any weight determined while the room is full of air is therefore inaccurate. The air in the room is holding up some of the weight of the bag. This mass loss is called **Buoyant Force**. It is calculated by multiplying the volume of the gas by the density of the air (the fluid). The mass indicated by the lab balance is called the **apparent mass** because it cannot include the mass upheld by the displaced fluid. The **true sample mass** of the sample is determined when the apparent mass is added to the buoyant force.

In this lab, a sample of propane gas was collected in a plastic bag. The bag was weighed to determine the apparent mass increase, caused by the gas in the bag. Next, the gas was transferred to an inverted graduate cylinder, which was filled with water. As the gas entered the cylinder, it displaced some of the water. The size of the gas bubble in the cylinder is read as the volume of the sample gas. This volume has two uses. It is first used to calculate the Buoyant Force. Next it is used to calculate the molar mass, with the **Ideal Gas Law**.

When Amadeo Avogadro hypothesized his concept of the mole, he stated that counting particles can be accomplished by dividing the sample mass by the atomic weight (molar mass).

In this lab, using the Ideal gas law, the number of moles was determined from the sample volume, the temperature, and pressure of the gas. The sample mass was also known, as that was a function of the apparent mass and the buoyant force. By applying basic algebra skills to Avogadro's formula and to the ideal gas law, one can solve for the molar mass of the gas sample. This becomes a very powerful tool in the science of analytical chemistry.

Title: Iodine clock reaction

Purpose:

All reactions rely on collision theory. Anything that supports a collision will speed up a chemical reaction. Reactions will run faster at higher *__concentration__*, higher **temperature**s, higher **pressures** (gases only) and with certain **catalysts**. In addition, the type of molecule, the type of bonds, the activation energy, and the shape of the molecule will also affect the rate. This experiment will test the concentration and temperature effects on reaction time. The two solutions are in an **Oxidation-Reduction** reaction. The Bisulfite solution has some **soluble starch** as an indicator material. When elemental Iodine is formed, the starch will turn colors.

Diagram:

Draw or attach a photo of the equipment in set-up mode

Procedure:

. Measure 20 ml of a *__0.02 Molar__* **Sodium Iodate** solution, as per the chart. Pour into a 400 ml beaker.

. Measure 20 ml of a *__0.06 Molar__* **Sodium Bisulfite/starch** solution, as per the chart. Pour this second solution into the first in the 400 ml beaker.

. Begin timing the reaction when the Bisulfite solution is poured into the Iodate solution in the beaker.

. Terminate timing at the moment a color change begins in the beaker.

. Record the time on the chart.

. Repeat steps 1-5 using the dilutions prescribed on the table.

PART II

. Make two tubes of trial #7 dilution strength, the slowest reaction time.

. Place the two solutions in a water bath **at 20-40-60-80** degrees.

. Add the two solutions to a beaker as you did in the first part of this experiment and begin timing.

. Repeat steps 7-9 for all temperatures listed in step 7.

42

DATA:

Complete the table

Calculations:

two step reaction.

$$(IO_3)^{-1} + 3\ (HSO_3)^{-1} \longrightarrow I^{-1} + 3(SO_4)^{-2} + 3\ H^{+1}$$

$$5\ I^{-1} + 6\ H^{+1} + (IO_3)^{-1} \longrightarrow 3\ I_2^{0} + 3\ H_2O$$

Make a graph of the Concentration/Time relationship

Make a second graph of the temperature / time relationship

Conclusion: Write an essay about this REDOX reaction.

Explain what is happening in each step of this reaction.

Explain (in detail) each of the various factors which can influence the rate of a chemical reaction.

Iodine Clock Reaction
A Study of reaction Rates

	IO_3	water	HSO_3	time	temp
trial 1	**20 ml**	0	**20 ml**		20
trial 2	**18 ml**	2 ml	**20 ml**		20
trial 3	**16 ml**	4 ml	**20 ml**		20
trial 4	**14 ml**	6 ml	**20 ml**		20
trial 5	**12 ml**	8 ml	**20 ml**		20
trial 6	**10 ml**	10 ml	**20 ml**		20
Trail 7	**8 ml**	12 ml	**20 ml**		**20**
	IO_3	water	HSO_3		
trial 8	8 ml	12 ml	20 ml		**40**
trial 9	8 ml	12 ml	20 ml		**60**
trial 10	8 ml	12 ml	20 ml		**80**

Solution A = 0.1 Molar IO3

Solution B = 0.1 Molar HSO3 + soluble starch

LAB #10

Chemical Reaction Rates

The rate or speed of a chemical reaction is defined by several factors. These factors are temperature, concentration, catalyst, and surface area. The complexity of the reaction is also a contributing factor. Any reaction requiring a complex series of molecular collisions will most likely run slower than a simple one-on-one reaction. In essence, reactions rely on collision theory. Anything that improves the number of force of a collision between particles will speed the reaction.

In this reaction, the temperature was kept a constant. The concentration of one reactant was also kept constant. The second reactant was subjected to serial dilutions to study the effect of concentration on the speed (rate) of the reaction. It was assumed that as the concentration decreased, the length of time for the reaction would increase, as the reaction would slow down. The collisions required for the completion of the reaction would take longer as less and materials were available.

In the procedure, we see that we began with equal quantities of the Sodium Iodate solution and the Sodium Bisulfite + Starch solution. As the experiment progressed, we recorded the elapsed time against the concentration. Each trial had two ml of the Iodate solution removed, as it was replaced with an equal volume of distilled water. In this way, the total volume of all materials in the test tubes was a constant. This also controlled the surface area of the reaction site. The results were then graphed as a function of time verses concentration. The graph clearly indicated that a geometric change is taking place.

A second investigation was undertaken using a fixed concentration of both materials. In this investigation, the temperature was increased in 20 degree increments, and the rate of the reaction was plotted against the temperature. Again we see a geometric correlation. As the temperature increased, the speed of the reaction also increased. It can be said therefore: ***As the temperature increased, the Time required for the reaction decreased.***

In conclusion, it is easy to state that the collision theory concept of chemical reactions is predictable. A catalyst was not used in this reaction, but it would have affected the rate of the reaction. The catalyst could have interfered with the collisions or in fact helped them. If one reactant is held in place by a catalyst, then the reaction speeds up. If the catalyst attaches to both reactants, then the reaction slows down. We saw that reducing the concentration had a slowing effect on the rate. Lowering the temperature also would slow a reaction. In our daily lives we see that "cold storage" helps to preserve many things from aging or "going bad". If energy is added to a reaction, it speeds up. If energy is removed, it slows down. This experiment confirmed several of the factors we had assumed would affect a chemical reaction. It also affirmed that concentrated material reacts rapidly, while diluted or weak materials react slowly.

LAB #11

Ideal Gas Law – Expected Volume of a Gas

Purpose:

Using a single replacement reaction between Hydrochloric Acid and pure Magnesium ribbon, predict the volume of hydrogen gas liberated.

Diagram:

Draw or attach a photo of the equipment in set-up mode

Procedure:

.   Carefully pour **30 ml of 3.0 molar HCL** into a 100 mL. gas collection buret.

.   Slowly add water onto the acid, taking care not to disturb the two layers.

.   Measure out **2.5 cm** of pure Magnesium ribbon, and it roll into a ball.

.   Weight this ball to four decimal places.

.   Wrap the ball with #22 gage copper wire to produce a circular cage with the Mg in the center.

.   Carefully insert and anchor the cage in the mouth of the burette.

.   Place your finger on the mouth of the burette, invert the buret, and place it into a beaker of water.

.   Clamp the burette in place.

. When the reaction has completed, determine the volume of gas collected.

. Compare this actual volume to the expected volume.

DATA:

Room Pressure **(P)** _____ torrs Vapor Press of water: _____

Dry gas pressure: _____ torrs   Atms: _____

Room Temp **(T)** _____ Celsius + 273 = _____ Kelvin

Mass of magnesium **(SM)** _____ grams

Molar Mass of magnesium **(MM)** = _____ **24.31** _____ grams/mole

Moles of magnesium used **(n)** = _____ moles

Expected Volume of hydrogen gas **(V)** = _____ liters = _____ ml

Actual volume of hydrogen gas collected = _____ ml

Percent Error = _____ %

Calculations:

Show the reaction using a Mass-Balance Map

Show all math used in proper HSPA format. Use 3 significant digits.

Conclusion:

Explain the Ideal Gas Law and the four gas law variables as they pertain to this lab.

Example Mass Balance:

Mg	+	2 HCl	$\longrightarrow$	MgCl$_2$	+	H$_2$
1 mole		2 moles		1 mole		1 mole
24.3 gr.						22.4 liters @ STP
0.0357 gr.						0.0353 liters

Mass of 1 meter of Mg ribbon = **1.428 g.**   Mass of 2.5 cm of Mg ribbon = **0.0357 g.**

n = SM / MM

n = 0.0357 / 24.4

n = 0.00147 moles

n = _ 0.00147 moles   **PV = nRT**

P = _ 760 /760 = 1 ATM   rm press   V = n R T / P

T = __23 C = 296 K   rm temp   V = 0.00147 * 0.0821 * 296 / 1

R = 0.0821 gas law constant   **V = 0.0352 liters = 35.3 ml**

V = _(35.3 ml)_ expected sample volume

V = _ (34.8 ml) _ actual sample volume
in buret

Percent Error:

Error = ABS difference / expected * 100%

Error = (35.3 – 34.8) / 35.3 * 100%

Error = 0.5 / 35.3 * 100%

Error = 1.42 %

The Four Gas Law Variables:
Temperature, Pressure, Volume, and Moles

All gases must be enclosed in a container that, if there are openings, can be sealed with no leaks. The three-dimensional space enclosed by the container walls is called **volume**. When the generalized variable of volume is discussed, the symbol V is used. Volume in chemistry is usually measured in liters (symbol = L) or milliliters (symbol = mL). A liter is also called a cubic decimeter (dm^3).

Other units of volume do occur such as cubic feet (cu. ft. or ft^3) or cubic centimeters (cc or cm^3). The main point to remember is: whatever units of volume are used, use them all the way through the problem. If you must convert from one unit to another, make sure you do it correctly. In the study and applications of chemistry, however, gas volumes are always expressed in liters.

Chemistry compares such things as Boyle's Law, Charle's Law and other laws where volume is a variable. That means the container involved in the experiment has a movable wall. When the volume goes up, that wall slides out. When the volume goes down, the wall slides in. This is VERY IMPORTANT to remember. Imagine the seal of the movable wall to be perfect so no gas escapes.

If the volume is constant, then the container is made with thick, rigid walls that cannot move. If the pressure increased too much, the walls would break, destroying the experiment. However, within the limits of any experiment discussed, the walls remain fixed and the volume stays constant. Pressure, volume, and temperature are in a tight relationship, called the Ideal Gas behavior.

All gases have a **temperature**, usually measured in degrees Celsius (symbol = °C). Note that Celsius is captalized since this was the name of a person (Anders Celsius). When the generalized variable of temperature is discussed, the symbol T is used. There is another temperature scale which is very important in gas behavior. It is called the Kelvin scale (symbol = K). Note that K does not have a degree sign and Kelvin is captalized because this was a person's title (Lord Kelvin: his given name was William Thomson). All gas law problems will be done with Kelvin temperatures. If you were to use degrees Celsius in any of your calculations, YOU WOULD BE WRONG! Kelvin temperatures are used because the slope of the graph between temperature and volume yields the gas law constant. You can convert between Celsius and Kelvin like this: Kelvin = Celsius + 273.15. Often, the value of 273 is used instead of 273.15. Check with your teacher on this point. All examples to follow will use 273. For example, 25 °C = 298 K, because 25 + 273 = 298. **Standard temperature is defined as zero degrees Celsius or 273 K.** The Kelvin temperature of a gas is directly proportional to its kinetic energy. Double the Kelvin temperature, you double the kinetic energy.

Gas **pressure** is created by the molecules of gas hitting the walls of the container. This concept is very important in helping you to understand gas behavior. Keep it solidly in mind. This idea of gas molecules hitting the wall will be used often. When the generalized variable of pressure is discussed, the symbol P is used.

There are three different units of pressure used in chemistry. This is an unfortunate situation, but we cannot change it. Different units are used for various concepts within the study of chemistry. These units are: atmospheres (symbol = atm), millimeters of mercury (symbol = mm Hg) and Pascals (symbol = Pa) or, more commonly, kiloPascals (symbol = kPa) **Standard pressure is defined as one atm. or 760.0 mm Hg or 101.325 kPa.**

Standard temperature and pressure is a very common phrase in chemistry, so common it has been abbreviated to STP. What the chemist will use as STP is actually called standard ambient temperature and pressure (STAP), but the difference between the two is unimportant in simple applications.

There is no such thing as standard **volume**, although molar volume is often used in studying gas behavior. The relationship of a certain mass of a sample gas to its volume, temperature and pressure was investigated by Amadeo Avogadro. He discovered that a specific quantity of gas always had the same volume at STP. The amount of gas is measured in moles (symbol = mol) or in grams (symbol = g or gm). This amount of gas occupies 22.4 liters at STP. When the amount of moles is discussed, the letter "n" is used as the symbol (note: the letter is in lowercase. (The other values have all been caps.). The **Ideal Gas Law** of Avogadro uses the **molar mass** of the sample gas and compares it to the actual **sample mass** of the gas sample being investigated. This yields the numbers of **moles** of gas in the system.

In this lab, a known sample of magnesium metal was reacted with a strong acid. The reaction produced a known quantity of hydrogen gas. The hydrogen was at room temperature and pressure. Applying the Ideal Gas Law allows one to calculate the expected volume of the sample gas formed by the reaction. The percent error then evaluates the method used to produce and collect the gas for its validity. As the error was low, the methods used in this experiment are valid.

LAB #12

Enthalpy change for a double replacement reaction

Purpose:

To determine the Enthalpy gain/loss when Magnesium Oxide reacts with Hydrochloric acid. If all reactants are in a proportion to the products, then the energy change of the system can be predicted.

Diagram:

Draw or attach a photo of the equipment in set-up mode

Procedure:

. weigh a clean, dry Styrofoam cup, (Calorimeter)

. Tare the balance and add **5.00 grams** of MgO to the cup.

. Obtain **100 ml of 6.0 Molar HCL** in a graduate cylinder [**Use extreme caution**]

. Using a glass thermometer, determine the starting temperature of the acid.

. *Wash off the thermometer* and then place into the calorimeter, on top of the magnesium oxide.

. Quickly, **BUT CAREFULLY**, pour the acid into the calorimeter. Slide the top over the cup.

. Stir rapidly until the temperature stops rising, *(swirling the cup).*

. Record the highest temperature of the solution.

. Wash everything down the sink and place thermometer back into holder.

. Calculate the Enthalpy change in this system. **WASH YOUR HANDS!**

Data:

Initial Mass of calorimeter, M1 _____gms

Final Mass of calorimeter, M2 _____ gms.

Mass of MgO in calorimeter _5.00_ gms

Initial Temp of Acid solution _____ Celsius

Final Temp of Acid solution _____ Celsius (delta T)

Temp Change of acid solution _____ Celsius

Total joules (work) released by system _____ joules (Q)

Expected Enthalpy change of system _141.2_ Kj/mole

Actual Enthalpy change of system _____ Kj/mole

Percent Error _____ %

Calculations:

Show the complete thermodynamic and the mass balance reactions

Show the determination of the actual heat gain/loss by this system **(Q)=M(T2-T1)Cp**

Show the determination of the **Enthalpy** and **Entropy** of this reaction.

Show the adjusted Enthalpy expectation of this reaction for the mass of MgO used.

Conclusion:

Explain this reaction on a step by step basis. Include what the **activation energy** is being used for and why energy is produced by this **Exothermic** reaction.

				Enthalpy	Entropy
$MgO \longrightarrow$	Mg	+	$\frac{1}{2} O_2$	+ 601.8	- 0.0268
$2\ HCl \longrightarrow$	H_2	+	Cl_2	+ 184.6	- 0.3734
$Mg + Cl_2 \longrightarrow$	$MgCl_2$			- 641.8	+ 0.0895
$H_2 + \frac{1}{2} O_2 \longrightarrow$	$H_2O_{(liq)}$			-286.0	+0.1398
$MgO + 2\ HCl \longrightarrow$	$MgCl_2$	+	H_2O	-141.4	- 0.1709

Mass of sample = 5.00 grams	$Q = M\ (\Delta T)\ Cp$
T1 water sample = 20 C.	Q = 100 g * 42 C. * 4.184 j/g-C.
T2 water sample = 62 C.	Q = 17,572.8 joules
ΔT water sample = 42 C.	Q = 17.6 Kj
Q = heat liberated by reaction	

Actual ΔH:

$$17.6\ Kj\ /\ 5.00\ grams = x\ /\ 40.3\ grams$$

$$x = 141.9\ Kj\ per\ mole\ of\ MgO$$

Error:

$$Error = abs\ difference\ /\ expected * 1005$$

$$Error = (141.9 - 141.4)/141.4 * 100\%$$

$$Error = 0.5\ /\ 141.4 * 100\ \%$$

$$Error = 0.354\ \%$$

Reverse Temp:

$$T = \Delta H\ /\ \Delta S$$

$$T = -141.4\ /\ -0.1709$$

$$T = 827\ Kelvin$$

Gibbs Free Heat:

$\Delta G = \Delta H - (T\Delta S)$

$\Delta G = -141.4 - (298 * - 0.1709)$

$\Delta G = - 192.3 \text{ Kj / mole}$ **(spontaneous)**

In this lab exercise, the concept of **Enthalpy** is investigated. Enthalpy is easily defined as the energy change during a chemical reaction. When investigating energy and its effect on a chemical system, the unit of **joule** is preferred over the **calorie** unit. The choice of the joule over the calorie is logical. A calorie is a unit of heat in the standard system. The joule is a unit of kinetic energy (work) in the metric system. When one analyzes the kinetic energy formula, we see **KE = 1/2 [mass * (velocity)^2]**

It is accepted that all matter has mass and all matter is in constant motion. The choice of a proper energy unit should therefore include these basic facts of matter. The joule is the quantity of energy associated with one kilogram moving at a velocity of one meter/second. By using the joule as the unit of energy, the investigator can determine the velocity of a particle of know mass at any temperature.

In this experiment, a sample of magnesium oxide was reacted with hydrochloric acid. Looking at the math of the reaction, one realizes that a simple double replacement reaction is actually four reactions. In reactions #1 and #2, the reacting materials have energy added to break their bonds so that the elements are liberated. Molecules are neutral, and atoms are reactive. The first two reactions have "activation energy" added to the molecules to break their bonds and return them to active atoms. In the third reaction, the liberated and activated magnesium from the MgO is now attracted to and reacts with the chlorine atoms. The chlorine was liberated in the decomposition reaction of the HCL into H_2 and Cl_2. The 4th and final reaction shows the hydrogen reacting with the liberated oxygen to form liquid water.

The first two reactions are **endothermic.** This means that energy is flowing into the reaction as "activation energy". The last two reactions are **exothermic**. They are liberating energy. Since the exothermic reactions produce more free energy than the first two reactions consume, the overall (net) reaction is exothermic.

Next, we look at the **Entropy** shift in the overall reaction. Entropy is the degree of randomness in a system. This reaction shows that the overall randomness had decreased as the reaction went on. In nature, the increase in entropy is favored. This reaction shows a decrease in entropy, which goes counter to the laws of thermodynamics. This sets the opportunity to reverse the overall reaction. To calculate whether a reaction can be reversed or not, one divides the enthalpy value (ΔH) by the entropy value (ΔS). The formula reads **(T = ΔH /ΔS)**. If the answer is positive, then the reaction can be reversed at the Kelvin temperature indicated by the answer. However, if the answer is negative, then the reaction cannot be reversed. **There are no negative Kelvin temperatures.**

Finally, we look at the **Gibbs free Heat, (ΔG)** of the reaction. This value determines whether the reaction is **spontaneous** or **non-spontaneous**. In addition, the reaction is rapid if the Gibbs value is greater than 200 Kj/mole and slow if less than 200Kj/mole. A negative ΔG indicates a spontaneous reaction. A positive ΔG is used for non-spontaneous reactions. As this overall reaction had a Gibbs Free Heat value of **(-192.3 Kj/mole)**, it is both spontaneous and relatively fast.

To summarize this reaction: the reaction is initiated by the addition of energy to decompose the molecular reactants into active, independent elements. These elements rearrange and react to form more stable, thermodynamically favored products. The overall reaction produces more heat than it consumes, therefore it is exothermic. The overall entropy has dropped. Since the sign of the Enthalpy value matches the sign of the entropy value, the reaction is reversible. Lastly, the Gibbs Free Heat value indicates that the overall reaction is both spontaneous (favored in nature) and rapid.

LAB #12

Enthalpy change for a double replacement reaction

Purpose:

To determine the Enthalpy gain/loss when Calcium Oxide reacts with Hydrochloric acid. If all reactants are in a proportion to the products, then the energy change of the system can be predicted.

Diagram:

Draw or attach a photo of the equipment in set-up mode

Procedure:

. weigh a clean, dry Styrofoam cup, (Calorimeter)

. Tare the balance and add **5.00 grams** of **CaO** to the cup.

. Obtain **100 ml of 6.0 Molar HCL** in a graduate cylinder **[Use extreme caution]**

. Using a glass thermometer, determine the starting temperature of the acid.

. *Wash off the thermometer* and then place into the calorimeter, on top of the calcium oxide.

. Quickly, **BUT CAREFULLY**, pour the acid into the calorimeter. Slide the top over the cup.

. Stir rapidly until the temperature stops rising, *(swirling the cup).*

. Record the highest temperature of the solution.

. Wash everything down the sink and place thermometer back into holder.

. Calculate the Enthalpy change in this system. **WASH YOUR HANDS!**

Data:

Initial Mass of calorimeter, M1 ⎯⎯⎯⎯⎯ grams

Final Mass of calorimeter, M2 ⎯⎯⎯⎯⎯ grams.

Mass of CaO in calorimeter _5.00_ grams

Mass of the 100 mo of HCl acid _100_ grams (**M**)

Initial Temp of Acid solution T_1 ⎯⎯⎯⎯⎯ Celsius

Final Temp of Acid solution T_2 ⎯⎯⎯⎯⎯ Celsius

Temp Change of acid solution ⎯⎯⎯⎯⎯ Celsius (**ΔH**)

Specific Heat Capacity of acid solution _4.184_ joules/gm-degree (**Cp**)

Total joules (work) released by system ⎯⎯⎯⎯⎯ joules (**Q**)

Expected Enthalpy change of system _260.9_ Kj/mole

Expected Enthalpy change of system ⎯⎯⎯⎯⎯ joules / 5 grams (**ΔH**)

Percent Yield ⎯⎯⎯⎯⎯ %

Calculations:

Show the complete thermodynamic and the mass balance reactions

Show the determination of the actual heat gain/loss by this system (**Q**) = **M(ΔT)Cp**

Show the calculations for **Enthalpy, Entropy, Gibbs Free Heat, reversing Temp**

Show the ***adjusted Enthalpy*** expectation of this reaction for the mass of **CaO** used.

The following is an example of the data analysis.

			ΔH	ΔS
$CaO \longrightarrow$	$Ca + \frac{1}{2} O_2$		+635.5	0.0397
$2\ HCl \longrightarrow$	$H_2 + Cl_2$		+184.6	- 0.3734
$Ca + Cl_2 \longrightarrow$	$CaCl_2$		- 795.0	+ 0.1138
$H_2 + \frac{1}{2} O_2 \longrightarrow$	$H_2O_{(liq)}$		-286.0	+0.0699
$CaCl_2 + 2\ HCl \longrightarrow CaCl_2 + H_2O_{(liq)}$			-260.9	- 0.2294

ENTHALPY : $\Delta H = - 260.9$ kj/mole ...... **Exothermic reaction**

ENTROPY: $\Delta S = - 0.2294$ kj/mole ... **Entropy increased**

-ΔH = exothermic   +ΔH = endothermic

Reversing Temp: T= $\Delta H /\Delta S$

T = - 260.9 / -.02294   ....   + 1137.3 ^{0} K. reaction <u>will reverse</u> at this temperature

Gibbs Free Heat -ΔG = spontaneous +ΔG = non-spontaneous

$\Delta G = \Delta H - (T\Delta S)$

ΔG = -260.9 – (298*(-.02294)

ΔG = - 260.9 + (68.36)

ΔG = - 192.54 kj/mole (Spontaneous, slow reaction)

Mass of acid = 100 g

T1 acid = 20

T2 acid = 76

ΔT acid = 56

Cp acid = 4.184 j / g-C

Q = ????

$$Q = M (\Delta T) Cp$$

$$Q = 100 \text{ g.}(56 \text{ }^0 C) \text{ } 4.184 \text{ j / g-C.}$$

$$Q = 23430.4 \text{ joules}$$

Mass of CaO =   **5.00 grams**

MM of Cao = 56.5 g/mole

Moles of CaO in sample = 0.089

$$\Delta H / 1 \text{ mole} = Q / N$$

$$260,900 \text{ joules / 1 mole} = Q / 0.089 \text{ moles}$$

Q = 23220.1 joules expected enthalpy

Percent Error = (difference / expected) * 100%

$$\% = (23,430.4 - 23,220.1) / 23,220.2 * 100\%$$

Error = 0.905 %

Conclusion:

Explain this reaction on a step by step basis. Include: **activation energy**, **spontaneous nature, reversing capability, Entropy change**

LAB #13

Determination of the Specific Heat Capacity of three pure metal samples

Purpose:

All substances require an amount of heat energy to change temperature. The quantity (Q) of heat which will change the temperature of one gram of mass, 1 degree Celsius is defined as the Specific Heat Capacity. The purpose of this exercise is to determine the Cp of Aluminum, Iron, and Copper.

Diagram:

Draw or attach a photo of the equipment in set-up mode

Procedure:

.   Weigh the three "pure" metal samples +/- 0.0001 grams

.   Set up the ring stand and beaker for boiling water

.   Bring to a boil 400 ml of water

.   Place each metal sample in the water and allow to heat until bubbles freely form on the surface of the samples.

.   Into a Styrofoam Calorimeter, place 100 ml of water

.   Record the temperature of this water (T1) using a digital thermometer.

.   Using ice tongs, remove a sample from the boiling water

. QUICKLY place it into the calorimeter.

. STIR RAPIDLY until the temperature stops increasing.

. Record the final temperature as T2.

. Clean out the calorimeter and repeat steps 5 – 10 with the other metal samples.

. Calculate the quantity of heat added to the cup (Q)

. Calculate the actual Cp of the metal.

. Calculate the percent error for each sample.

Data:

	Al	Fe	Cu
Mass of water in cup	100 g	100 g	100 g
Mass of sample (M)			
T1 water in cup			
T2 water in cup			
Delta T of water			
Heat gained by cup (Qw)			
T1 metal	100 C	100 C	100 C
T2 metal			
Delta T metal			
Heat lost by metal (Qm)			
Actual Cp			
Expected Cp	0.90	0.45	0.39
% error			

Calculations:

Show all math required to complete the table, in standard format

Example data:

Aluminum sample

Mass = 18.05 g.

T_1 sample = 100 C.

T_2 sample = 23 C.

ΔT sample = 77 C.

Water in calorimeter

Mass = 100 g.

T_1 water in cup = 20 C.

T_2 water in cup = 23 C.

ΔT water in cup = 3 C.

Joules added to water = mass * ΔT * Cp

Q = 100 g. * 3 C. * 4.184 j/g.-C.

Q = 1255.2 joules

Note: joules added to the water was lost by the metal as it cooled

Cp of metal sample = Q / M* ΔT

Cp = 1255.2 joules / 18.05 g. * 77 C.

Cp =   0.903 j / g.-C.

Error = (ABS difference / expected) * 100%

Error = (0.903 – 0.90) / 0.9 * 100%

Error = 0.003 / 0.9 * 100%

Error = 0.333 %

Repeat the above steps for all three samples

Specific Heat Capacity

Matter is divided into different classes. There are the metals which have **Luster, Malleability, Ductility**, and **Conductivity**. There are the non-metals, which are **Brittle, Dull**, and act like **Insulators**. Finally, there are the Metalloids, which share some of the properties of both metals and non-metals. These properties are directly linked to the molar mass, atomic size, atomic spacing, and density of the material under investigation. In addition, metals tend to donate electrons during a chemical reaction, whereas non-metals tend to gain the electrons. All of these properties and behaviors contribute to the ability of a material to transfer heat throughout the sample. In this lab exercise, these topics will be investigated. The goal is to determine why a substance has a Specific Heat Capacity unique to that material. Specific heat Capacity is defined as the quantity of energy required to cause a one degree Celsius change in one gram of a substance. The energy is measured in units called joules. A joule is a unit of energy derived from a force applied through a distance. **A joule is a Kilogram-meter^2/Second^2**

In the 1890's there were two scientists, **DuLong and Petit**, who performed experiments on various pure materials (elements) to compare their **Specific Heat Capacity** to the **Atomic Number** and also to the Density. They stumbled onto a relationship that as the Atomic Number increased, the Specific Heat Capacity decreased. Early into their experiments, it was determined that there were many factors influencing the density of a material. This made an accurate comparison to the Specific Heat Capacity impossible. These two men then graphed the results of the experimentation. It became obvious that if the Specific Heat Capacity was known, then the Atomic Number could be accurately surmised. They had discovered another tool in the creation of the science known today as **<u>Analytical Chemistry</u>**.

When the atomic number increases, there is also an increase in the atomic mass. This increase is caused by the addition of protons, electrons, and neutrons to the atom. The increasing sub-atomic particle count brings with it an increase in the **mass – to - charge ratio** of that element. *The increasing attractive forces within the atom, causes a <u>reduction</u> in the atomic volume.* In other words, the radius of the heavier atoms shrink, due to these increasing attractive forces. The neighboring atoms are now closer to each other. Orbiting electrons are closer to their nucleus. As the atoms draw closer, they have a greater ability to transfer energy from one to another. The <u>empty space between the atoms was acting like</u> <u>insulation.</u> Therefore, the transfer of energy was blocked and larger amounts were required to "heat" the neighboring atoms. As the space was reduced, the insulating property of that space was also being reduced. Energy given to atoms near the surface was now more easily transferred to the inner atoms. As a result, less energy was needed to make a temperature change in the material. The transfer of energy became more and more efficient. The overall Specific Heat Capacity was reduced. Space between atoms is a place where energy is lost. Anything that reduces these inter-atomic spaces will improve the ability of that material to conduct heat. As a material gets hotter, its atoms spread away from each other, the intra-atomic spacing increases. The conductivity of this material is now decreased. Conversely, as a material is cooled, these spaces shrink, and its ability to conduct will improve. If the material was to be **Super-Cooled**, then it becomes a **Super-Conductor**.

LAB #14

Melting Point behavior of a pure substance

Purpose:

To determine the freezing point depression of a pure substance. In this experiment, three trials will be performed. In each trial, the mass of solute is increased in 2.0 gram increments. The mass of solvent remains constant at 30 grams.

Diagram:

Draw or attach a photo of the equipment in set-up mode

Procedure:

.   Set up a double boiler on a ring stand

.   Obtain a large (25 X 150 mm) or larger test tube

.   Place **30 grams** of **Para-dichloro-benzene** (PDCB) crystals into the tube. This is the solvent.

.   Place a thermometer into the tube and lower entire apparatus into the boiling water.

.   When the crystals are all melted, and the temperature reads 65 Celsius, remove from the water.

.   Allow the crystals to air cool and record the temperature every 30 seconds.

.   Take 12 – 16 readings (6 – 8 minutes), ***after the temperature plateaus***.

.   Record the freezing point temperature.

. Lower the tube back into the boiling water to melt the sample.

. Add **2.0 grams** of <u>**Napthalein**</u> crystals to the sample and stir.

. When the sample again gets to 65 Celsius, remove from the bath.

. Record the temperature every 30 seconds until the new freezing point is determined.

. Repeat steps 10, 11, and 12 two more times so that a total of 6.0 grams of the Napthalein has been added to the PDCB.

. Determine the freezing point depression from the data for each solution.

Data:

collect temperature reading at 30 second intervals. Make a minimum of 30 readings.

Calculations:

Show the graph of the data collected.

The FP depression is a constant. Show your proof in the math used to solve for the molar mass.

Can you calculate the molar mass of the solute using this data?

Conclusion:

When a material is added to a pure substance, the freezing point is lowered. Explain the process of freezing.

Explain why a solute affects the freezing point by lowering it.

SOLUTE:

SM of Napthalein: T_1 = 2.0 grams T_2 = 4.0 grams T_3 = 6.0 grams

SOLVENT:

Mass of Paradichlorobenzene: _____30 grams …….. **(0.03 Kg)**

Initial MP/FP of pure PDCB : _____**53**_Celsius

T_1 New MP/FP:	___49__C. (2.0 grams of solute added)	ΔMP = **4.0 C.**
T_2 new MP/FP:	___45__C. (4.0 grams of solute added)	ΔMP = **4.0 C.**
T_3 new MP/FP:	___41.5__C. (6.0 grams of solute added)	ΔMP = **3.5 C.**

Average ΔMP for three trials: _____**3.83**_____ **Celsius per 2.0 grams of solute**

K_f of PDCB = 7.1 Celsius/mole-Kg

MM of solute = (SM * K_f) / (solvent mass * ΔT)

MM of solute = (2.0 g. * 7.1 C/mole-Kg) / (0.03Kg * 3.83 C)

MM of solute = 123.6 g / mole …… **experimental answer: 124 g / mole**

Napthalein formula.. $C_{10}H_8$ …………. **actual molar mass: 128 g / mole**

Error = [Absolute difference / expected] * 100%   *[units not shown to simplify]*

Error = [128 – 124] / 128 * 100%

Error = 4 / 128 * 100%

Error = **3.12%**

Freezing Point Depression – lab 14

At a given pressure, the temperature at which a specific *pure* substance undergoes a particular phase change (freezing, boiling, sublimation, etc.) is always the same. For example, pure water at 1 ATM of pressure will always freeze at 0°C and boil at 100°C. However, if impurities (solutes) are added to the pure solvent in the liquid phase, the temperatures at which phase changes from the liquid occur will be different than those for the pure solvent.

The freezing point of a solution formed by mixing a pure liquid with a small amount of solute will decrease and the boiling point of the solution will increase. These effects are referred to as ***freezing point depression*** and ***boiling point elevation.*** These changes are part of a larger class of related phenomena called **colligative properties.** Other examples include the lowering of the vapor pressure of a pure liquid by the addition of solutes and the phenomena of osmotic pressure, which is important in keeping blood cells from collapsing.

In our everyday lives, the phenomenon of freezing point depression explains why road ice will melt when salt is poured on it (the freezing point is lowered by the addition of the impurity) and why we put ethylene glycol (antifreeze) in our automobile radiators to protect our cooling systems from freezing in extreme cold. Physically, freezing point depression and boiling point elevation occur because a solution consisting of a solvent (the formerly pure liquid) with a small amount of solute added is thermodynamically more stable than the pure liquid itself - primarily due to the fact that the addition of the solute **increases the entropy** of the solution. This additional stability increases the range of temperatures at which the liquid is stable - *lowering* the freezing point and *increasing* the boiling point. In this lab you will be exploring how the freezing point depression of water depends upon the amount and nature of the added solute.

The new freezing point occurs when the solution has been cooled to a temperature where the solute begins to settle out of solution. The solvent is now free of the interfering solute, and can begin to freeze. As the solvent freezes, the remaining solvent becomes more concentrated with the solute which had been originally added. This new, more concentrated solution, therefore has a still lower freezing point. When one inspects the graph in this lab, you see a flat line for the pure solvent. The solutions however all contain a sloping line, showing a lowering of the FP over time. This is evidence of the increasing solute concentration is the diminishing (liquid) solvent mass.

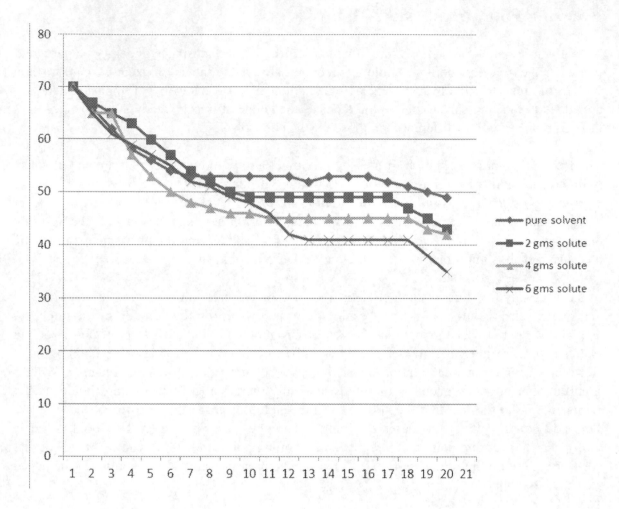

LAB #15

Building a simple voltaic cell

Purpose:

Any device using a series of chemical reactions to liberate electrons can be called a battery. This lab has two salt solutions and two metal electrodes creating electricity.

Diagram:

draw and label the cell (**quarter page maximum size**)

Procedure:

. Clean and weigh a **zinc electrode** (Anode)

. Set up two 400 ml beakers.

. In one beaker, place 300 ml of 3.0 M. **copper sulfate** solution

. In the other beaker, place 300 ml of a 3.0 M **zinc sulfate** solution.

. Build a **salt bridge** and insert between the two beakers above.

. Place a clean **copper electrode** in the copper sulfate solution. (Cathode)

. Using a **RED** wire, connect the **copper electrode** to the voltmeter.

. Using a **BLACK** wire, connect the **zinc electrode** to the meter.

. Place the zinc electrode into the zinc sulfate solution and time it for exactly **15 minutes.**

. At the end of this time, remove the Zn electrode, clean with steel wool, wash it off, and dry it.

. Reweigh the Zn electrode to determine the mass loss.

DATA:

Mass of Zn (I)	_____ grams
Mass of Zn (F)	_____ grams
Mass loss on Zn	_____ grams
Molar mass of Zinc	_____ grams/mole
Time cell ran	__ 15 min__ = __900__ seconds

Recorded voltage	_____ volts	
Expected **voltage**	__1.104_____ volts (V) or (E)	
Calculated **Amps**	_____ coulombs/second (I)	
Calculated **Resistance**	_____ Ohms (R)	**Resistance = Volts / Amps**
Power production of this cell	_____ Watts (P)	**Watts = Volts * Amps**
Heat produced by the voltage	_____ Joules (Q)	**Joules = watts * time**
Gibbs Free Heat (**spontaneous**)	_____ Kj/mole	**delta G = - nFE^0**
Life expectancy of this cell	_____ Hours	
Temp change on **600 ml** of battery solution	_____ Celsius	

Calculations:

Show the complete REDOX reaction

Show all math in HSPA format

Conclusion:

Explain how this battery produces electricity. Explain the role of each component in the process. Explain the equilibrium requirements of the two half cells using the **Nernst Equation**, **Le Chatelier's Principle**, and the need for the **salt bridge**. Explain the changes in the system when two voltaic cells are connected in **Series** and then in **Parallel**.

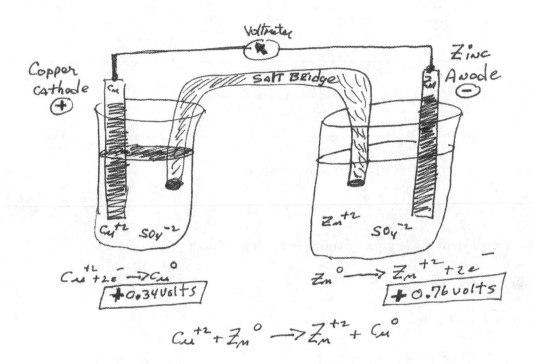

Zinc ion	Copper ion	Q	log Q	Voltage	change
0.1	0.1	1.00	0.000	**1.1000**	NA
0.2	0.1	2.00	0.301	1.0911	**-0.0089**
0.3	0.1	3.00	0.477	1.0859	**-0.0141**
0.4	0.1	4.00	0.602	1.0822	**-0.0178**
0.5	0.1	5.00	0.699	1.0793	**-0.0207**
0.1	**0.1**	1.00	0.000	**1.1000**	NA
0.1	**0.2**	0.50	-0.301	1.1089	**0.0089**
0.1	**0.3**	0.33	-0.478	1.1141	**0.0141**
0.1	**0.4**	0.25	-0.602	1.1178	**0.0178**
0.1	**0.5**	0.20	-0.699	1.1206	**0.0207**

$E^0 = E - (0.059/2) \log Q$

$Q = [Zn^{+2}] / [Cu^{+2}]$

$E^0 = 1.104$ volts

Zn^0 $+ Cu^{+2}$ ------- Cu^0 $+ Zn^{+2}$ **1.104 volts**

Amps = (SM/MM) X charge X 96,500 coul X 1/ time

Delta G = - n*F*E⁰   **F = 96,500 coulombs/mole of e⁻**

$$E^0 = 1.104 \text{ volts}$$

Gibb's Free Heat = (-1)(moles oxidized)(96,500)(voltage)

Power = Volts X Amps

(watts) = volts * amps

Joules = Watts X time

Joules = (Mass) * (Temp change) * (Spec. Heat)

$$\text{Spec Heat of } H_2O = 4.184 \text{ joules/gram-}^0C$$

Temp Change = Joules / (mass of solutions)*(Cp)

The formation of any series of chemical reactions which can **liberate electrons** is called a **battery**. Batteries similar to the type made in this lab exercise have been around for centuries. In the 1840's, a scientist named Michael Faraday discovered that electrons flow from an active metal to a less active material. Faraday determined that when electrons flow on a wire, they create a **magnetic field.**

In this **electrochemical cell**, the zinc was the more active metal. It has an **electronegativity** of **[1.65]**. The copper is less active because it has an electronegativity of **[1.90]**. The higher value for copper demonstrates that it would prefer to gain electrons, whereas the lower value for the zinc proves that it prefers to lose electrons. (Materials which lose electrons easily are called **active metals**. Materials which lose electrons are said to be **oxidizing.**)

The **thermodynamics** of the two major materials in the battery also points toward zinc as the likely material to oxidize. The standard heat of formation (**Enthalpy**) for Zinc Sulfate is **-983 Kj/mole**. The standard heat of formation for Copper Sulfate is less than the zinc salt with a value of **-771 Kj/mole**. Since nature prefers reactions having the greater enthalpy, the zinc sulfate is favored over the formation of copper sulfate. Both of these materials are **electrolytes**: therefore, they must **dissociate** in water. Since zinc has the greater enthalpy value, it will dissociate easily.

In turn, it also gives up its electrons easily. Zinc can readily transition from the insoluble **atomic state** into the more soluble **ionic state** better than copper can.

Electrons are sub-atomic particles in motion. All particles have mass. All particles are in constant motion. Mass times Motion gives us Kinetic Energy. This kinetic energy is measured as the force created by these moving particles. This force is called **voltage.** Any material which liberates electrons to freely travel along a wire is a source of electricity. The more electrons liberated, or the greater the force they are pushed with, the greater the voltage. **Oxidization is electrons being pushed <u>out of</u> a material. Reduction is electrons being <u>pulled into</u> a material.**

Electrons which flow from one material into another can be diverted onto a path outside of the reaction vessel. This path is called a **circuit**. This external circuit allows the use the electrons while they are "en route" to their final destination, the material being reduced. The electrons carry a charge of -1. This charge is fantastically small. The unit of this charge is measured in **Coulombs**. If one considers the number of electrons liberated by one mole of atoms undergoing oxidation, then the charge can be formidable. The total charge on one mole of electrons was determined by Michael Faraday to be **96, 500 coulombs**. If we couple the total charge in coulombs to the time it takes for the cell to run, then we get coulombs / second. A coulomb flowing per one second is called an **Ampere.**

The simplest way to look at amperes is to consider them to be the number of workers in a circuit. Volts could be regarded as the force that is driving these workers. Finally, the **total Power** in the system **(WATTS)** can be viewed as the total work being performed by the total number of workers (AMPS) in one second of time. Watts are calculated by getting the product of volts and amps. Many electrical appliances are measured in watts. A toaster for instance, consumes 900 watts. A hair dryer consumes 1500 watts. Most people are familiar with the term wattage from the sizing of light bulbs. The higher the wattage of the bulb, the brighter and hotter the bulb will be. A night-light, for example, is only 7 watts.

As with all systems doing work, there is a consideration of energy loss in the system. The **Ohm** is a unit of electrical **resistance.** As the resistance increases, the loss of voltage in the circuit also increases. Engineers use a variety of in-line resistors to build electrical devices. The resistor gives control over the voltage, which in turn, allows the engineer to custom design the device. As the resistance builds, the voltage drops, and the heat builds up.

When **thermochemistry** was investigated, the term joules was first introduced. A **joule** is a unit of work. This work is often expressed as heat. Work is a unit of **Kinetic Energy**. If resistance (ohms) can produce heat, then ohms are related to joules. Therefore, we have a connection between Joules (Q) and Ohms (R), and Watts (w.)

Gibbs Free Heat is a measure of the spontaneous nature of a chemical reaction. It is derived from the balance of the enthalpy, temperature, and entropy of a chemical system. When ΔG is **negative**, the **reaction is exothermic and spontaneous**. This system may be called a **battery**. When ΔG is positive, the **reaction is endothermic and non-spontaneous**. This latter type of system may be called **electrolysis**. A battery is a series of spontaneous chemical reactions which produces free electrons to power electrical devices. The concept of cause and effect has come full circle. The type of bonds in the reactants has determined: the nature of the chemical reaction, the quantity and type of energy released by that reaction, and the overall activity of the reactants. The more ionic the bonds, the more likely that oxidation will occur. The more covalent the bonds, the more likely that reduction will occur.

In closing, batteries are nothing more than a series of interacting reactants exchanging electrons. The exchange liberates electrons from one material as they are consumed by another. In many similar reactions, energy in the form of heat is associated with these processes. Electrochemistry points out that the energy released could be **electricity**. The same laws of nature apply regardless of the energy form.

LAB #16

Determination of the solubility of a compound in 100 ml of distilled water, at a constant temperature

Purpose:

To calculate the mass of salt which can dissolve in 100 grams of water. You will then compare the molar masses of all samples tested. You will also draw a proposed shape of each molecule tested. An internet search will help you with this part of the assignment. And finally, you will determine the type of bonds present using electronegativity data, (percent ionic character).

Diagram:

Draw or attach a photo of the equipment in set-up mode

Procedure:

.   Measure 100 ml of distilled water in a graduated cylinder. **Record the temperature.**

.   Place water into the 250 ml beaker provided.

.   Obtain a plastic cup containing the sample to be tested.

.   Weigh the cup and sample material to within 0.001 grams.

.   Slowly using the spatula provided, add the sample to the water while stirring constantly.   [*If an electronic stirrer is available, use it*].

.   When additional sample will no longer dissolve into solution, reweigh the plastic cup.

.   Calculate the missing mass of solute from the cup.

. The missing mass is the solubility per 100 ml. Multiply by 10 to determine the solubility per liter (1,000 ml).

. Repeat steps 1 – 8 for all samples

. Compare you findings to data available in reference materials and the internet.

Data:

	NaCl	CaCl$_2$	FeCl$_3$	Pb(NO$_3$)$_2$
Molar Mass of solute	58.35 g/M.	111.2g/M	162.2 g/M	331.0 g/M
Mass of cup, initial	_____	_____	_____	_____
Mass of cup, final	_____	_____	_____	_____
Mass of sample used	_____	_____	_____	_____
Soluble mass / 1000 ml	_____	_____	_____	_____

Conclusion:

. Which sample was the most soluble?

. Which sample was the least soluble?

. Which samples did you expect to be most and least soluble?

. Why do you think the sample which was the most soluble had this property? Explain each reason.

. Why do you think the sample which was the least soluble had this property? Explain each reason.

. What roll would the temperature play in solubility? Explain your answer.

. How do you expect table sugar (sucrose) to behave? Actually test table sugar and compare to your hypothesis. How did your hypothesis make out compared to the actual results for the sugar? *Sucrose has a molar mass of 342 g/M.*

LAB #17

Combustion of methyl alcohol

Purpose:

The combustion of this alcohol releases great quantities of heat. In this experiment, you will determine the Enthalpy, Entropy, and Gibbs Free Heat.

In addition, you will also determine if the reaction can reverse.

Diagram:

Draw or attach a photo of the equipment in set-up mode

Procedure:

.   Weigh the aluminum **calorimeter***

.   Weigh the alcohol lamp

.   Place **100 ml** of water into the calorimeter and set up the **apparatus***

.   Light the alcohol lamp.

.   Heat the calorimeter water until a **10 degree** temperature change is achieved.

.   Cover the lamp and quickly reweigh it.

.   Calculate the mass of fuel used to heat the system 10 degrees.

.   Calculate the total heat placed into the system by the flame (Qt)

.   Compare to the Enthalpy expected.

James Signorelli

DATA:

Mass of Aluminum calorimeter: _____ grams (M$_C$)

Mass of water in calorimeter _100_ grams (Mw)

Mass of lamp, initial: _____ grams

Mass of lamp, final: _____ grams

Mass of fuel burned (consumed) _____ grams moles of fuel consumed _____

T1 (initial temp - water in cup) _____ Celsius

T2 (final temp – water in cup) _____ Celsius

ΔT (temp change of water in cup) _ 10 _ Celsius

Heat added to calorimeter: _____ joules (Qcup)

Heat added to water: _____ joules (Qwater)

Total heat generated: _____ joules (Qc + Qw)

Total Enthalpy change in system: _____ Kj

Expected Enthalpy change for mass _____ Kj
of fuel used :

Actual Enthalpy change in cup: _____ Kj

% error: _____ %

Calculations:

Show the Mass Balance

Show the calculations for Enthalpy, Entropy, and Gibbs Heat.

Show the calculation for the reversing temperature.

Show the calculation for the percent error

Conclusion:

Explain this combustion reaction using Hess's law. What happens at each step, to both the thermodynamics and the Entropy.

EXAMPLE OF MATH

$Q = M (\Delta T) Cp$

Calorimeter
$Q = M_C (\Delta T)Cp$
$Q = \textbf{78.23g}\ (10C)\ 0.90\ j/g\text{-}C.$
$Q = 704.07$ joules

Water
$Q = M_W(\Delta T)Cp$
$Q = \textbf{100g}\ (10\ C)\ 4.184\ j/g\text{-}C.$
$Q = 4184$ joules.

Total heat added to system = 4888.07 joules

$Q_{total} = 4.888$ kjoules

Moles of Methanol consumed = _____ grams.

Actual Enthalpy change per one mole _____ Kjoules

$2\ CH_3OH + 3\ O_2 \longrightarrow 2\ CO_2\ +\ 4\ H_2O$

		Enthalpy	Entropy
$2\ CH_3OH \longrightarrow$	$2\ C + 4\ H_2$	$\Delta H =$	$\Delta S =$
$3\ O_2 \longrightarrow$	$6\ O$	$\Delta H = 0$	$\Delta S =$
$2\ C + 2\ O_2 \longrightarrow$	$2\ CO_2$	$\Delta H =$	$\Delta S =$
$4\ H_2 + O_2 \longrightarrow$	$4\ H_2O$	$\Delta H =$	$\Delta S =$

$2\ CH_3OH + 3\ O_2 \longrightarrow 2\ CO_2\ +\ 4\ H_2O$

Enthalpy change =
Entropy change =

Gibbs Free Heat
$\Delta G = \Delta H - (T\Delta S)$

- **The apparatus is an aluminum calorimeter positioned on a wire triangle. The triangle is resting on top of a metal can with both the top and bottom removed. This can is resting on three rubber stoppers, to create the updraft. The alcohol lamp is inside of the entire apparatus.**

LAB #18

Calculating Absolute Zero

Purpose:

The determination of absolute zero by William Thompson (aka: Lord Kelvin) was never performed at extreme temperatures. The pressure exerted by a gas sample sealed in a metal sphere was all that was needed. The behavior of any gas is linear to the temperature, until a phase change occurs.

Diagram:

Draw or attach a photo of the equipment in set-up mode

Procedure:

. Set up a large four liter beaker OR a. eight quart metal soup pot on a hot plate.

. Place ice into the beaker (pot) and the Pressure / Temperature apparatus.

. Record the pressure of the sealed gas after the system reaches thermoequilibrium (0 C.)

. Turn on the hot plate to highest setting.

. Record the temperature & pressure every ten degrees until the water begins to boil.

. Graph the data (Temp: y-axis vs Press: x-axis)

. To determine absolute zero, plot the "Y" intercept of your graph.

DATA:

T = 0	C.	P = _____	***pressure may be expressed in any unit.***
T = 10	C.	P = _____	
T = 20	C.	P = _____	
T = 30	C.	P = _____	
T = 40	C.	P = _____	
T = 50	C.	P = _____	
T = 60	C.	P = _____	
T = 70	C.	P = _____	
T = 80	C.	P = _____	
T = 90	C.	P = _____	
T = 100	C.	P = _____	

Calculations:

Show the graph using a symbol around the data point. Use a ruler to draw the line. A solid line for the data actually collected and a dotted line for the interpolated data. Use this line to determine the "y" intercept. The intercept is the proposed **absolute zero**, a temperature where the pressure would be reduced to zero. This graph assumes that no phase change occurs as the gas cools. This can't happen in the real world, so it is assumed to be an **"Ideal Gas"** which the graph is representing.

Conclusion:

Research the work of William Thompson and write an essay on the importance of Absolute Zero.

The concept of absolute zero has intrigued man for almost 150 years. As any real gas is heated, the kinetic energy gained by the molecules increases their velocity. The more energetic molecules now hit the walls of their container more often and with a greater force. If the container is elastic (movable walls), the increasing collisions and force of those collisions pushes against the walls. The walls in an ***elastic container*** will move in response to the increasing attack by the energized molecules. So for an elastic container, the **increase in temperature causes an increase in the volume of that gas**. This is a reflection of Charles' law.

However, if the container has ***rigid walls***, then a different change occurs. As the molecule velocity is increased by an elevating temperature, the walls can't move out. **In response to the increasing number of collisions and the increasing force of those collisions, the pressure begins to rise**. The changes are linear and directly related to the increase in the temperature. This is now a reflection of the law of Joseph Gay-Lussac.

The pressure units used for this exercise are not important, as the change is directly related to the changing temperature. Any pressure unit is acceptable. Some prefer atmospheres of pressure. Others prefer mm of Hg (aka: torrs). Still others use millibars or kilopascals. What is important is that degrees Celsius are used for the temperature. We can't use degrees Kelvin as yet, because the purpose of this lab is to discover the conversion value to create degrees Kelvin. That conversion value is the (y-intercept) added to the degrees Celsius.

The data collected clearly shows that a ten degree increase in temperature causes a constant increase in the pressure. As the pressure bulb used by this researcher was measured in pounds per square inch, the ten degree increase caused a 0.5 PSI increase. Our range of temperature was from 0 Celsius to 100 Celsius. The range of pressures as a result was from 14 PSI (0 C.)to 19 PSI.(100 C). Laying a ruler on the data points crates a straight line **(constant slope).** If we now extend the line down to the "Y" axis, we see that the line touches at approx. – 275 Celsius. The accepted value for absolute zero is -273.15 Celsius.

This lab used equipment costing very little. With that in mind, the actual value for absolute zero was determined to be with less than 1% of error. In the 1850's, when absolute zero was first tested, an animal bladder connected to a tube, and mercury manometer was used. While our equipment was less involved, our results were the same. In science, a law will express itself, regardless of the equipment used.

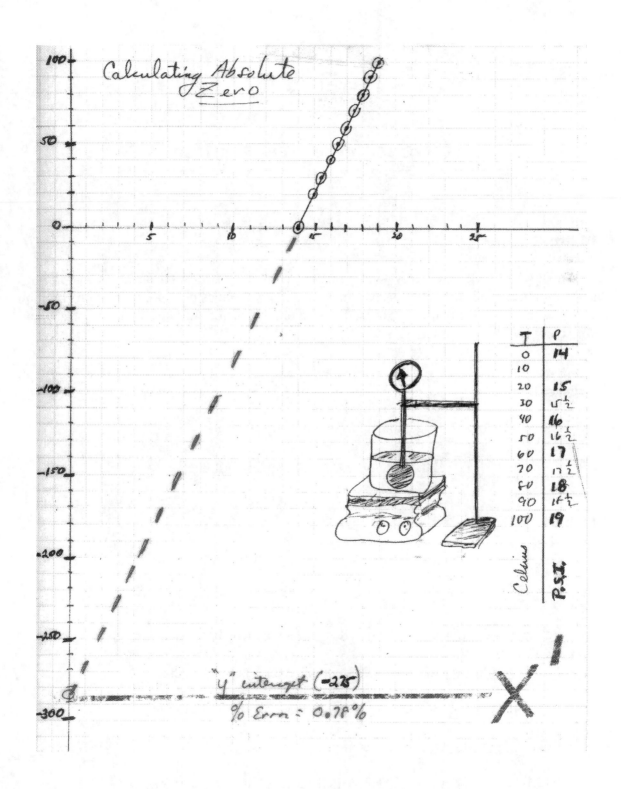

Calculating Absolute Zero

T	P
0	14
10	
20	15
30	5½
40	16
50	16½
60	17
70	17½
80	18
90	18½
100	19
Celsius	P.S.I.

"y" intercept (−275)

% Error = 0.78%

LAB #19

Titration of an acid solution

Purpose:

Acetic acid (CH_3COOH) is a very weak acid. The dissociation constant for this solute implies that the number of ions formed is a small fraction of the molecules placed into solution. This lab exercise will investigate the Ka, [H+], % dissociation, and pH of a dilute solution of acetic acid *(white vinegar)*.

Diagram:

Draw or attach a photo of the equipment in set-up mode

Put your lab aprons and goggles on now

Procedure:

. Obtain a clean 400 ml beaker.

. Place a small stirring bar into the beaker.

. Set up a titration device. (two burettes) and a magnetic hot plate.

. Fill the **left burette** with the acid solution (vinegar) to be tested

. Using a pH meter, measure the acid. Calculate the [H+] concentration.

. Fill the **right burette** with the control solution. (NaOH)

. REMEMBER TO PURGE THE AIR IN THE BURETTE VALVE

. Dispense 20 drops of the *Phenylphlthalein* indicator solution into the beaker

. Dispense 20 mL of the acid solution into the beaker with the indicator.

. Slowly dispense the base solution into the beaker until a pale pink (blush) color is discharged. ***Slowly stir to avoid splashing the base solution.***

. Add one drop of acid to the beaker. If the color pink returns to clear, then you have found the **"end-point"**.

. Record the volume of base used to neutralize the acid solution. Do this three times and **determine the average volume of base from the three trials.**

. Calculate the Molarity, expected [H+], the pH, and % dissociation of the acid using the correct formula for each.

DATA:

example of data collected from prior experiment

pH meter reading: —————— 2.42 ——————   **[H+] = _0.0038_moles/liter** (from pH)

predicted Molarity (from pH meter data) _0.802_moles/liter

Volume of Base   ————————— mL (average of three trials)
Molarity of Base   ____ 0.50 ————————— moles/liter
Volume of Acid   ____ 20 —————————— mL
Molarity of Acid   ————————— moles/liter (calculated from titration)
Ka Acetic acid   ___**1.8 x 10^{-5}** —————————
[H+] expected   ————————— moles/liter
pH expected   —————————
% dissociation   ————————— %

James Signorelli

Reaction:

HAc + NaOH ⟶ **Na^{+1}+   Ac^{-1} + H$_2$O**

CH$_3$COOH is abbreviated as HAc   (CH$_3$COO)$^{-1}$ is abbreviated as Ac^{-1}

Example Calculations: **[use standard math format]**

Show all math required to complete the data table

T1: _33.0_ ml   total volume = 99.8 ml

T2: _32.6_ ml   **average volume of base:** 99.8 ml / 3 trials = __**33.3**__ml

T3: _34.2_ ml

MaVa = MbVb   **actual Ma from pH**

                                                           Ma = [H]^2 / Ka

Ma = MbVb/Ma   **Ma = [.0038]^2 / 1.8 x 10^{-5}**

                                                           Ma = 0.8022 M./liter

Ma = (0.5 m/l) * **33.3 ml** / 20 ml   **Ma = 0.802 M./liter.. (3 sig. digits)**

Ma = 0.833 moles / liter

H$^+$ = √ (Ma * Ka)   **actual from pH: H$^+$ = 10^{-pH}**

H$^+$ = √ **(0.833 M. / l * 1.8 X 10^{-5})**   **H$^+$ = 10$^{-2.42}$**

H$^+$ = 0.00387 M. / l   **H$^+$ = 0.0038 M.**

---

pH = -log (H$^+$)

pH = -log (0.00387)

pH = 2.41

Error = [absolute difference / expected] * 100%   (**based on Molarity values**)

Error = [0.833 – 0.802 / 0.802] * 100%

Error = 0.031 / 0.802 * 100 %

Error = 3.87 %

% Dissoc = [H⁺] / M. * 100%

% dissoc = 0.00387 / 0.833 * 100%

% dissoc = 0.465 %

LAB # 19

A **titration** is a form of a **double replacement** reaction. The French verb **'titre'** means literally "to compare". The reaction of an acid with a base forms a salt solution and water. This reaction is **spontaneous** and **exothermic**. The acid / base interaction causes the solutions to **neutralize** each other. Therefore, a titration can also be called a neutralization reaction.

All acids, bases, and salts are classified as **"electrolytes".** That implies that in solution, these materials do not exist as intact molecules, but rather as **dissociated** ions. Ionic solutions have very definite properties. Among these properties is the ability to conduct electricity. The pH meter functions because the ions in solution conduct electricity in the meter. The conductivity is related to the ion concentration of the solution. The tip of the pH meter has a glass bulb with microscopic holes in the glass. These holes only allow the passage of a specific size ion into the detector bulb. Therefore, the detector is a specific ion meter. The term "meter" means "to measure". So our pH meter is specific for measuring only the ions which can pass through the holes and conduct electricity in that meter.

In this reaction, a simple distilled, white vinegar solution was obtained from a supermarket. The label on the vinegar lists that the concentration is 4% acidity. This value implies that 96% of this solution is water, and the remaining 4% are acetic acid molecules. The water content is variable in other vinegar solutions (cider, wine, balsamic, etc.) as flavor and color molecules are also present. To be labeled as vinegar, the FDA requires that the acidity is fixed at 4%. Therefore, it was necessary that the white vinegar was used for this experiment as the water and acid content was a known value.

The first formula used in this experiment was MaVa = MbVb. If the formula is analyzed using the math units, we see that the Molarity multiplied by the volume leave moles of material. This formula compares the moles of acid against the moles of base. It also compares the volumes required to create this balance from solutions of various concentrations. The connection of the formulae used in the lab requires a specific order of solving. 1st: Determine the average volume of base required to neutralize the acid. 2nd:Determine the moles of acid molecules in one liter of that solution. 3rd: Determine the concentration of hydronium ions created by the Molarity of acid molecules. 4th: Determine the pH using the hydronium ion concentration. 5th: Determine the percent dissociation from the data collected in the experiment.

A net ionic reaction must clearly indicate that the molecules of electrolytes do not combine as long as water is present. This neutralization reaction forms liquid water as one of the products, and the reactants are both solutions. The acetic acid is really the dissociated **hydronium** and **acetate ions**. The base is composed of the dissociated sodium and **hydroxyl ions**. The sodium ion cannot combine with the acetate ion as water is present. The hydronium ion does combine with the hydroxyl ion, forming the liquid water. This __net ionic reaction__ is depicted as:

$$H^+_{aq} + (CH_3COO)^{-1}_{aq} + \quad Na^+_{(aq)} + (OH)^{-1}_{aq} \longrightarrow Na^+_{aq} + (CH_3COO)^{-1}_{aq} + H_2O$$

Rarely does all of the electrolyte molecules dissociate into ions. The actually ratio of intact molecules to dissociated ions creates a value called the dissociation constant (Ka). The percentage of molecules in the original solution (before dissociation) compared to the actual concentration of a specific ion is the percent dissociation. While the percent dissociation can change as the Molarity of the original solution changes, the Ka remains a constant. The only way to change the Ka is to change the temperature. As the temperature is increased, more molecules of the electrolyte will dissociate, therefore, shifting (increasing) the KA value.

A titration is a very powerful tool in the analytical chemistry lab. It allows the determination of concentration of the test material. Additional experiments can determine the identity of that material. Combining the data from various analytical tests can be used to completely identify the substance and its concentration. This method of analysis is used in medicine, forensic science, petro-chemistry, organic chemistry, biochemistry, and geochemistry. In the hands of a well trained chemist, the titration method of materials solves many unknowns.

LAB #20

Single Replacement Reaction of Iron and Copper Sulfate

Purpose:

The oxidation number of an element determines the reaction ratio between the metal and the non-metal. In this experiment, iron filings will be reacted with a warm copper sulfate solution. The warm solution speeds the action so that all required activities may be completed in 60 minutes.

Diagram:

Draw or attach a photo of the equipment in set-up mode

Procedure:

Day #1

.   Weigh **2.00 grams** of fine iron filings to two at least decimal places.

.   Obtain **100 ml of a 1.0 Molar Copper Sulfate solution** in a clean 400 ml beaker

.   Warm the solution to at least 80 degrees Celsius.

.   ***SLOWLY add the iron filings to the heated solution***. This reaction is ***EXOTHERMIC*** and care must be taken not to have it boil over from the additional heat produced by the reaction.

.   Allow the reaction to run for 10 minutes, keeping the temperature between 80 – 90 degrees Celsius. Reheat as needed.

.   Set-up a filtering apparatus.

. Weigh a clean piece of filter paper to decimal places (or finer).

. Filter the contents of the beaker to reclaim the precipitate, **(ppt.).**

. Use a wash bottle to transfer all of the copper onto the paper in the funnel.

. Wash the **(ppt.)** on the filter paper several times to remove all un-reacted copper sulfate.

. Set aside to dry overnight.

. **<u>WASH YOUR HANDS</u>** with soap & water before leaving the room!

DAY #2

. Reweigh the filter paper and determine the mass of precipitate.

. Calculate the moles of Copper liberated by this reaction.

. Calculate the combining ratio of the Iron & Copper

DATA:

Mass of iron filings _____grams

Moles of iron used _____moles

Mass of filter paper, day #1 _____grams

Mass of filter paper, day #2 _____grams

Mass of copper precipitate _____grams

Moles of copper liberated _____moles

Molar ratio _____

Calculations:

Show the possible chemical reaction.

Show all math associated with this lab, in standard computational format.

Conclusion:

The iron replaced the copper in the **compound.** The **reaction** had iron giving electrons to the copper to make it release the sulfate. The release allowed the iron to now attach to the sulfate. The copper settles out as a **precipitate. <u>Explain in detail.</u>**

LAB #20

A **single replacement reaction** exchanges a pure element for a similar element in a compound. In this lab, iron, the pure element, is reacting with a copper salt solution. The iron and copper are both metals, so it is expected that the more active metal will replace the least active metal. Inspection of the periodic table lists iron as having two possible **oxidation states**. One state has iron giving up two electrons in a reaction. The other state has it donating three electrons in a reaction. As a part of this lab exercise, the oxidation state exhibited by the iron will be investigated. The combining ration of the iron to the copper sulfate is our clue. If one iron replaces one copper, than both must have the same oxidation state. It is known that the sulfate ion has a value of -2. Since our copper is bonded 1:1 with the sulfate, then the copper must have a state of +2.

The reaction is heated to initiate the process. Although it is an exothermic reaction, the warm solution used in this experiment speeds the process. Care is taken not to allow the solution to reach the boiling point as the process liberates extra heat from the exothermic reaction. If the solution boils over onto the lab table, weighing the product becomes impossible as some of the copper will have been ejected from the beaker. When a reaction takes place, there are indicators of an on-going process. For some reactions, it is the release of a gas. For other reactions, it could be a temperature change. This reaction cause the release of gas bubbles as the electron pass from the iron into the copper. The electrons cause some of the water to decompose into its component gases. You will hear a hissing sound if you carefully place the beaker near your ears. This is proof of a gas being discharged. The reaction also causes the solution to get hotter and then begin to boil. This is why care must be taken to remove the external heat source and prevent the solution from boiling over. In addition, during many chemical reactions, a color change is evident. Your solution began as a blue liquid. As the reaction progressed, much of the blue color was lost. The new soluble product, iron sulfate, is actually yellow in color. This will be observed during the filtering process.

After the reaction has run its course, the copper product (**precipitate**) is filtered and washed. The washing is to remove excess copper sulfate and the by-product which is iron sulfate. The original filter paper was white. If after filtering, the paper is either blue or yellow, then continue washing. The excess copper sulfate and iron sulfate by-products are staining the paper. You must was the filter and precipitate until the paper is back to its original color (white). The precipitate is than set aside to **dry overnight**. If dried in an oven, the copper may react with the oxygen in the air to form an oxide, which would make the product mass heavier than expected.

Lab note: The solutions used in this lab could be bad for the environment. Follow your instructor's directions as to the proper disposal of the filtered solutions. Never dispose of liquids down the sink drain unless the instructor gives the direction to do so. Often, a liquid waste container is used for these waste solutions.

LAB #21

Observing the reaction rate of metals with water

Purpose:

To determine the cause of different reaction rates for various metals as they react with water. When metals react with water, a basic solution is formed. The water will be pre-treated with a base indicator called **phenylphthalein.** This indicator turns pink in the presence of a base.

As the pH approaches 14, the indicator turns completely red. (looks like CoolAid)

Diagram:

Draw or attach a photo of the equipment in set-up mode

Procedure:

.   Obtain three 250 ml beakers. Place a ***clean piece of white paper*** under these beakers.

.   Place exactly **100 grams** (100 ml) of water into each beaker

.   Add five drops of the indicator solution to each beaker.

.   Drop in the premeasured metal sample, a different sample into each of the three beakers. (Mg, Ca, Zn)

.   Observe the metal in the water for several minutes. Record any observations which indicate that a reaction may be occurring.

.   Observe the color of the solution every minute. Keep a minute by minute log of these observations.

Conclusion:

Atomic radius is a measure of the size of the atom. Valence electrons are farther away from the nucleus in larger atoms.

Atoms with a weak hold on their electron react faster than atoms with a stronger hold on the electrons. Research the electronegativity of the three sample metals.

Many metals react with water to form a base. (exceptions: Au, Pt, SS)

Compare the reaction time of three metals to their atomic radius and electronegativity. Make a chart showing each value.

Make a conclusion which relates the data in your table to the reaction performance.

LAB #22

Esterification

Purpose:

An alcohol and a carboxylic acid join by dehydration-synthesis to form an oily substance called an ester. This ester if found in plants is called an oil. If found in animals, it is called a fat.

Diagram:

Draw or attach a photo of the equipment in set-up mode

Procedure:

.   place 5 ml of the alcohol into a large test tube

.   place 3 ml of the organic acid into this same test tube

.   place sulfuric acid catalyst into this mixture created in steps #1 & #2

.   heat gently over a low flame moving the tube back and forth through the flame.

.   when the aroma of the ester is detected, stop heating.

.   allow mixture to cool and waft your hand over the mouth of the tube.

.   record the appearance and aroma of the product.

Components	product	aroma
A. Ethyl alcohol & Acetic Acid	ethyl acetate	fruity
		(nail polish remover)
B. Ethyl alcohol & Butyric Acid	ethyl butyrate	Pineapple
C. Iso-Pentyl alcohol & Acetic Acid	iso-amyl acetate	Banana
D. Methyl Alcohol & Salicylic Acid	methyl salicylate	Wintergreen
E. Methyl Alcohol & Butyric Acid	methyl butyrate	Green Apple
F. Ethyl Alcohol & Lauric Acid	ethyl laurate	Pear

Observation:

Conclusion:

. Draw each reaction including the formation of the liquid water from the dehydration step.

. Explain the process called Dehydration-Synthesis (condensation), including the Thermodynamics of the reaction. (These are all spontaneous reactions)

Ester Lab

The formation of large organic molecules usually involves a reaction mechanism called **Dehydration-Synthesis.** When several smaller organic molecules are joined, they have been prepared by heating in the presence of a strong, **hydroscopic** catalyst. In living systems, these catalysts are called **enzymes**. In the lab, the material of choice is sulfuric acid. This acid is so hydroscopic that the addition of it to the reaction vessel begins the reaction.

The joining of an alcohol to an organic acid produces a substance known as an **Ester.** The joining requires that the alcohol gives up the hydroxyl (-OH) group, and that the organic acid loses the hydrogen (H) from its carboxyl group. The loss of these two fragments creates charged molecules which immediately snap together into the ester. The ester would gain back the water formed if that water was not "locked up" by the hydroscopic catalyst. The reversal reaction would be called **Hydrolysis**. Any material made by Dehydration-Synthesis carries a label: **Store in a cool, dark, dry place.** The cool and dark is to deny the reaction any activation energy. The dry is to also deny the reaction one of the reactants for the reversal direction.

Esters are defined as oils and fats. The oils tend to be smaller, plant based materials with a characteristic aroma and flavor. Oils are liquids at room temperature. Fats tend to be large, animal based esters, also carrying an aroma and flavor. Fats are solids at room temperature. Esters give most of our foods their flavor and fragrance. The flavor and fragrance industry, world-wide, is a 900 billion dollar business. These companies deal with aroma and flavor oils, mainly esters.

The formation of an ester is **Exothermic**. Therefore, it is also **spontaneous** and favored in nature. The reversal process (hydrolysis) requires the addition of energy to proceed. That explains the cool and dark requirement for storage. Because esters can decompose (reverse) into the original material, they have a shelf-life. Esters made from large fatty acids are easily transformed into **cholesterol**. These esters can be also changed into **trans fats** when their multiple bonds are broken and hydrogen is added back.

The most important concept to be discussed in this lab is that none of the materials formed, nor any of the reactants came from a living source. All of the materials and components were artificial, petroleum based chemicals. Man has conquered nature in that we no longer must harvest materials after growing or raising them. No plants or animal must be sacrificed to provide for the supply of these flavor or fragrance items. We can simply make them in the lab. *The artificial materials are identical to the natural chemicals with several exceptions: We can make them 100% pure; we can make them in vast quantities; and we can make it all overnight.*

Esters

Ethyl Alcohol + Acetic Acid $\xrightarrow{H_2SO_4}$ Ethyl Acetate + H_2O

Ethyl Alcohol + Butyric Acid $\xrightarrow{H_2SO_4}$ Ethyl Butyrate (Pineapple Oil)

Iso-Amyl Alcohol (Iso-Pentyl) + Acetic Acid $\xrightarrow{H_2SO_4}$ Iso-Amyl Acetate (Banana Oil)

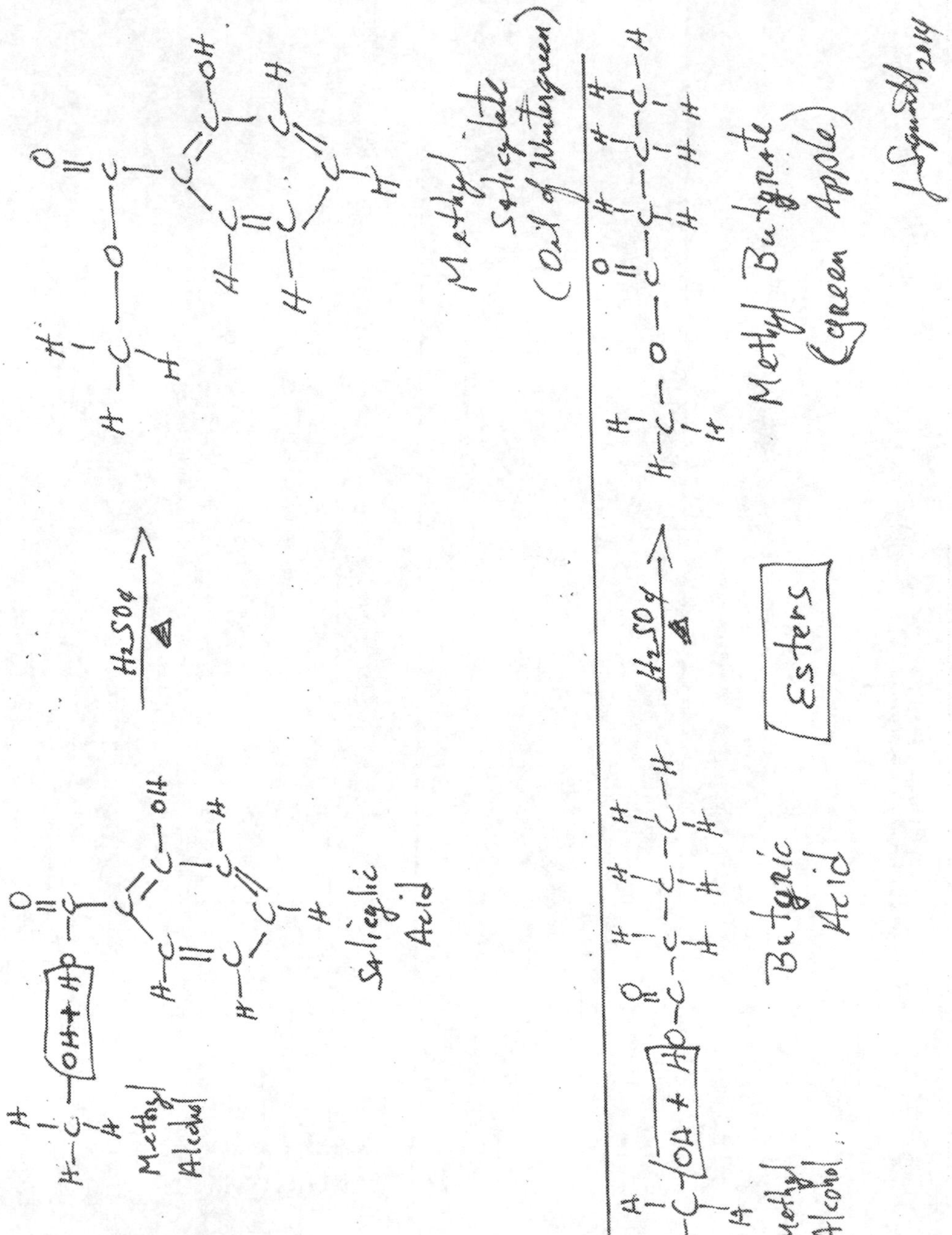

Methyl Alcohol + Salicylic Acid →(H₂SO₄)→ Methyl Salicylate (Oil of Wintergreen)

Methyl Alcohol + Butyric Acid →(H₂SO₄)→ Methyl Butyrate (Green Apple)

Esters

J Signorelli 2014

LAB #23

The Hydrolysis of an Ester

Purpose:

All fats and plant oils are esters. When an ester is reacted with a strong base, Hydrolysis will split the ester into an alcohol and the salt of the ester. If a triglyceride is the ester, than the alcohol formed is glycerol.

Diagram:

Draw or attach a photo of the equipment in set-up mode

Place you safety glasses on now. Put on your apron.

Procedure:

. Place 10 mL of oil (corn, olive, animal fat, etc.) into a 600 ml beaker

. Add 10 NaOH pellets

. Add 5 mL of a NaCl solution (the catalyst)

. Heat until boiling on a hot plate. Reduce the heat to medium.

. Stir constantly and rapidly using a long glass rod..

. When the soap becomes creamy,(resembles mashed potatoes), **SLOWLY** add a few ml of 3.0 M. NaOH solution to the beaker.

. Stir until a crumbly solid is formed.

. Allow the solid to cool on paper towels. Use a metal spatula to scrap out the beaker.

. Take two small pellets and place one each into a small test tube.

. In one tube, add 5 ml of tap water. To the other tube, add 5 ml of distilled water. Shake rapidly and observe each tube carefully.

. Take a small pellet of soap and wash your hands with it. How do your hands feel?

Data:

Draw the entire structural reaction

Observations:

Conclusion:

Explain the process of Hydrolysis and how it is related to Saponification

Give a brief history of Soap and Soap making

Explain how soap works. Why does it clean? Describe the **Hydrophobic** and **Hydrophilic** properties of soap.

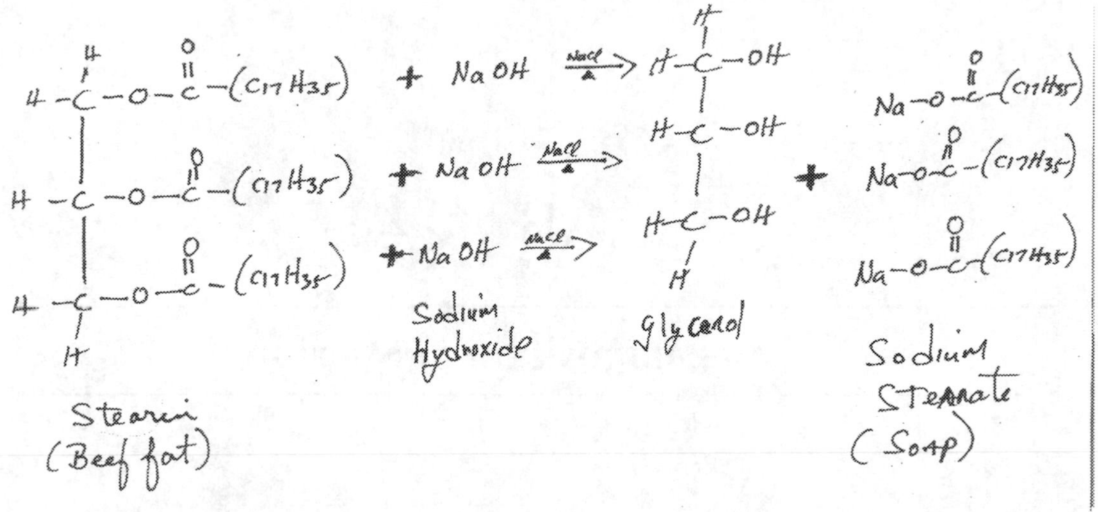

Stearin
(Beef fat)

$+ \ NaOH \xrightarrow{NaCl}$

Sodium
Hydroxide

Glycerol

Sodium
Stearate
(Soap)

SOAP

LAB #24

Decomposition of Sucrose

Purpose:

This lab will have the student decompose sucrose by extreme heating in a Bunsen burner flame. The products of this reaction are activated charcoal (pure carbon) and water vapor.

Diagram:

draw the equipment as it was set up or attach a photo

Procedure:

- Weigh a small plastic cup. **Set balance 2.00 grams heavier.**

- Weigh 2.00 grams of sucrose in the cup.

- Weigh a small test tube.

- Place the sugar into the tube.

- Light your Bunsen burner or alcohol lamp.

- Place the clamp onto the test tube.

- Slowly heat the tube & sugar until it begins to melt.

- Pass the tube back and forth in the flame and continue to heat.

- When the sample stops smoking, it is done.

- Cool the tube before laying it on the table.

. Turn off the burner or lamp.

. Reweigh the tube and determine the mass of carbon remaining in the tube.

DATA:

Mass of sucrose	_2.00_ grams
Mass of empty tube	_____ grams (A)
Mass of tube + sample	_____ grams (before heating)
Mass of tube + carbon	_____ grams (after heating) (B)
Mass of carbon (actual)	_____ grams (B-A)
Expected mass of carbon	_0.84_ grams
Percent Composition	_____ % carbon

Calculations:

Show all mathematics in standard format

Conclusion:

Explain what a decomposition reaction is. Explain why the carbon filled the tube, and the sugar sample only occupied a small space before heating. What are the various stages the sugar passes through as you heated it?

LAB #25

Acetylation of Salicylic Acid: "Aspirin" production

Purpose:

To investigate a substitution reaction. In addition, to determine the history and chemistry of a "wonder drug" called Aspirin. **[Acetyl-Salicylic Acid]**

Diagram:

Draw or attach a photo of the equipment in set-up mode

Procedure:

note the order of addition for steps 1-2-3

. Carefully pour 3 ml of 6.0 molar Acetic Anhydride into a large test tube

. slowly add 20 drops of conc. (18 M.) Sulfuric Acid (dehydration catalyst)

. Measure out 5.0 gm of pure Salicylic Acid crystals and add to the test tube

. Heat tube gently in a bath of hot water at 90 C.

. Heat mixture for ***five minutes*** or until an oily film develops on the surface of the liquid.

. Cool off mixture by running cold water along the outside of the tube. If it condenses into a solid, then you may filter and wash the sample. If not, reheat for two more minutes.

. Set up the filtering apparatus and wash the filtrate several times to remove the sulfuric acid and any un-reacted acetic anhydride. **Use well chilled, distilled water.**

. After the water drains from the filtrate, feel to residue. It should feel slippery and slimy.

Alternative to #7 & #8: rinse product in tube and scoop out the Aspirin.

Conclusion:

(Each lab group could report on two of the selected eight topics)

. Explain the history of Aspirin, beginning with ancient civilizations using willow bark, and ending with **Felix Hoffman**, the inventor.

. Aspirin is an NSAID : **Non-Steroid Anti-Inflammatory Drug. Explain this term.**

. Explain how Aspirin works. Give a brief report on **Sir John Vane**, the winner of the1982 Nobel Prize in Medicine for this discovery.

. What is **Reye's Syndrome**?

. Why should pregnant women and hemophiliacs avoid aspirin?

. What is **Kawasaki's Disease**?

. Recent research shows a link to Aspirin and the ***prevention*** of **Colorectal Cancer**. Prepare a brief report on this research.

. **Explain these key terms associated with aspirin:**

Analgesic

Anti-pyretic

Anti-coagulant

Vasodilator

Antiseptic or Antibiotic

Anti-Inflammatory

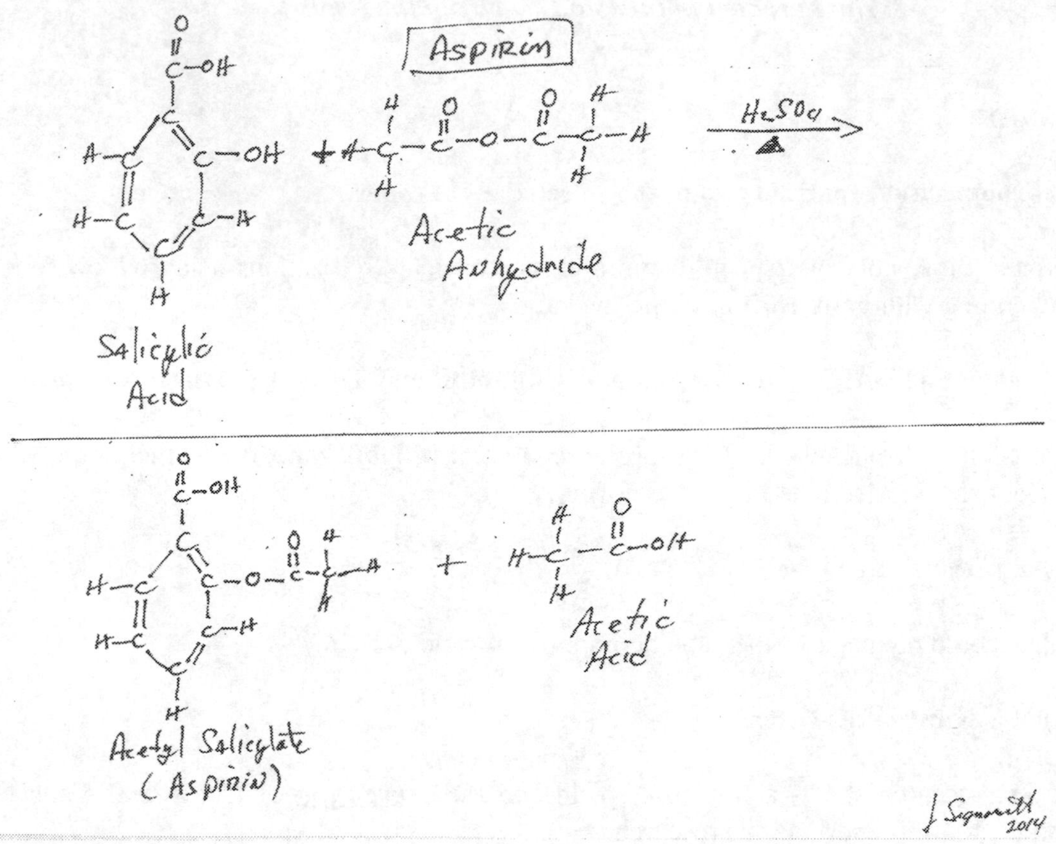

James Signorelli

Aspirin

Salicylic
Acid

Acetic
Anhydride

Acetyl Salicylate
(Aspirin)

Acetic
Acid

J Signorelli
2014

110

LAB #26

AP level
Synthesis of Bakelite

Introduction

Chemists classify polymers in several ways. There are thermosetting (insoluble, does not soften at high temperatures) plastics such as Bakelite and melamine and the much larger category of thermoplastic materials, which can be molded, blown, and formed after polymerization. There are arbitrary distinctions made among plastics, elastomers, and fibers. And there are the two broad categories formed by the polymerization reaction itself: (1) addition polymers (e.g. vinyl polymerization), in which a double bond of a monomer is transformed into a single bond between monomers, and (2) condensation polymers (e.g.Bakelite, nylon), in which a small molecule, such as water or alcohol is split out as the polymerization reaction occurs.

In this experiment, Bakelite is prepared by condensation of phenol and formaldehyde in a base catalyzed reaction. Like other thermosetting polymers, Bakelite is not soluble in any solvent and does not soften when heated. It is the plastic used to make the black handles on kitchen pots and pans that withstand the heat of an oven.

Diagram:

Draw or attach a photo of the equipment in set-up mode

Procedure

In a 25-mL round-bottomed flask place 3.0 g of phenol and 10 mL of 37% by weight aqueous formaldehyde solution. The formaldehyde solution contains 10-15% methanol, which has been added as a stabilizer to prevent the polymerization of formaldehyde.

Add 1.5 mL of concentrated ammonium hydroxide to the solution and reflux it for 5 min beyond the point at which the solution turns cloudy, a total reflux time of about 10 min.

In the hood pour the warm solution into a disposable test tube and draw off (decant) the upper layer. Immediately clean the flask with a small amount of acetone.

Warm the viscous milky lower layer on the steam bath and add acetic acid drop wise with thorough mixing (shaking) until the layer is clear, even when the polymer is cooled to room temperature.

Heat the tube on a water bath at 60-65 Celsius for 30 min.

Then, after placing a wood stick in the polymer to use as a handle, leave the tube in an 85 Celsius oven overnight. To free the polymer, the tube may need to be broken. Attach a piece of the polymer to your lab report.

Conclusion:

. **Draw the structure of this heat setting resin and explain the reaction.**

. **Write a short history of the discovery and uses for Bakelite.**

LAB #27

The preparation of a polymer, Thiokol, a synthetic rubber

Purpose:

To create a polymer compound in the lab from paint solvent and powdered sulfur,

Diagram:

Draw or attach a photo of the equipment in set-up mode

Procedure:

.   In a 250 ml beaker, boil 2 grams of sulfur with 10 grams of NaOH, dissolved in 150 ml distilled water.

.   Boil until the solution turns Amber color.

.   Remove from heat, allow to settle, and decant the liquid and save.

.   Return the amber liquid (sodium sulfide solution) to the hotplate.

.   Bring solution to 72 degrees and hold at this temperature.

.   Slowly add 10 ml of Ethylene Dichloride to the solution, stirring constantly. If a stirring hot plate is available, use it. Set stir to slow.

.   Heat gently and stir for 10 minutes until the solution turns "milky".

.   Stir for an additional 5 minutes

.   Remove from heat and slowly add 2 ml of glacial Acetic Acid.

. Stir for two minutes and let stand.

. Decant liquid into a WASTE container.

. Wash rubber disc under a stream of water while squeezing it into a ball shape.

. Remove from the wash and bounce ball on table.

Observations:

Record your observations about the appearance and smell at each stage of this reaction.

Conclusion:

. Draw the entire reaction as it progressed from step one through step 10

. Write a brief history of rubber, both natural and synthetic. Include the names of the giants in the industry. (Harvey Firestone, Charles Goodyear, etc)

LAB #28

Determination of the diameter of a wire using density data

Purpose:

A wire is an elongated cylinder. The metal the wire is composed of is known to be 99.99% pure. Using simple mathematical equations, determine the volume, radius, and diameter of the wire. This lab points out the step by step procedures a scientist can follow to gather facts about a sample. *It also points out that accurate science was possible long before high technology was invented!*

Diagram:

Draw or attach a photo of the equipment in set-up mode

Procedure:

.   Measure a 50 cm piece of copper wire.

.   Wipe the wire with an alcohol pad to remove all oils and fingerprints.

.   Weigh the wire to four decimal places.

.   Using the density and sample mass data, calculate the volume of the sample.

.   Using the volume and length data, calculate the radius of the sample.

.   Using the calculated radius, predict the diameter of the sample.

.   Using a micrometer, measure the diameter of the wire.

. Calculate the percent error.

DATA:

Mass of sample:	_____ grams	
Length of the wire:	_____50_____ cm	*(Height)*
Density of the wire:	_____8.96____ grams/cc	*(pure copper)*
Calculated volume of the wire:	_____ cc	*(volume)*
Calculated radius of the wire:	_____ cm	*solving for (r)*
Expected diameter of wire:	_____ cm	
Actual diameter of the wire:	_____ cm	*(micrometer reading)*
% error:	_____ %	

Calculations:

Show all math required to complete the table, in standard math format

$$Dn = Mass / Vol \qquad Vol = Mass / Dn \qquad Vol = (3.14) * (R)^2 * H$$

Conclusion:

Explain: Why did the wire require cleaning before weighing? What assumption is made about the uniformity of this wire? What could be a source or error in this method?

LAB #29

Measuring the Surface Tension of a Liquid

Purpose:

using capillary tubes, one can measure the surface tension of a liquid as a function of the radius of the tube and the height of the liquid.

Diagram:

Draw or attach a photo of the equipment in set-up mode

Procedure:

. obtain a capillary tube.

. to measure the inside diameter, slide a fine wire into the bore. Make sure that the wire selected has a tight fit.

. Using a micrometer, measure the diameter of the wire. This is the internal diameter of the tube.

. Next, measure the height the liquid rises in the tube above the base fluid level.

. Record the temperature of the fluid being tested.

. Calculate the surface tension using the charts provided and the data collected.

DATA:

Wire diameter: _____ mm   _____ cm

Wire radius: **(r)**   _____ mm   _____ cm

Fluid height **(H)**   _____ mm   _____ cm

Acceleration of gravity **(g)**   **_980.616_** cm/sec^2

Temperature   _____ ^{0}Celsius

Density of Fluid **(p)**   _____ gm/cm^3

Calculated surface tension (☐)   _____ dynes/cm

Expected surface tension (☐)   _____ dynes / cm

% error   _____ %

Calculations:

Show the math associated with the solution in standard math format

$$☐ = ½ (p*g*r*H) \qquad\qquad g = 980.616 \text{ cm/sec}^2$$

Conclusion:

write a complete analysis of surface tension, adhesions and cohesion.

(Remember: as (r) decreases, (H) increases)

Surface Tension

A liquid is composed of particles (atoms or molecules) which interact with each other. They exist is a phase where by the particles move somewhat freely about their container. This liquid phase can be defined to have **no definite shape**, but due to internal forces, it **has a definite volume**. This lab investigated one of those forces, called **surface tension**. This force is a result of a balance between adhesion and cohesion. **Adhesion** is a property of matter whereby a material (adheres) sticks to another substance it is in contact with. **Cohesion** is a property where particles of a specific substance (cohere) stick to their neighboring particles, of that same substance. This can be simplified to say: cohesion – likes stick to likes; adhesion – opposite materials stick to opposite materials.

The fluid under investigation also has properties which can influence the adhesion and cohesion forces. If the molecule is polar, then is would have stronger attractive forces than a neutral, non-polar substance. Large massive particles have a stronger attractive force for each other due to Newton's laws of gravitational attraction. Charged particles can have a very strong attractive force on nearby opposite charges. If the substance under investigation has a strong **Dipole**, it will also have a strong Surface tension. If no Dipole is present, that it clearly is expected to have a weak surface tension.

Water has a significant surface tension. It is caused by water molecules **cohering** (attaching) to each other. As you recall from basic chemistry, water is a **Polar Molecule**. As the temperature changes, this force can be increased or decreased. Just before reaching the freezing point, this force becomes so strong, that it is renamed "**Hydrogen Bonding**". When water is confined in a narrow tube, adhesion of water molecules for the walls of the tube causes the water to creep up the sides of the tube. This adhesion of the water for the walls of the tube creates a condition known as **capillary action.**

Since this force must defy gravity to move up the walls, the ***acceleration of gravity*** (g) becomes a key term in the formula to calculate the surface tension. Next, the diameter of the tube is considered. As the walls become closer with the narrowing of the tube, the force easily "grabs" and "pulls" itself up. So the radius of the tube is also in the formula to calculate surface tension. Next, the density of the liquid in the tube must be known. It is a known fact that density is a function of temperature. As the temperature increases, the volume of a liquid increases. Therefore, the density of that liquid decreases. So now we must add the ***density of the liquid (p)*** in the tube to the formula. Finally, the ***height (H)*** that the fluid has climbed in the tube must be inserted into the formula. This height is a function of how well the fluid has defied gravity and pulled itself up the walls of the tube.

How does water get from the roots to the leaves in a tall tree? It is the surface tension and capillary action which carries the water up the tubes of the tree. This living system is a function of the physics and chemistry at work in the bark, where the capillary tubes are located. The study of fluids involves the math and application of these concepts associated with the cause of the property called surface tension.

LAB #30

Analytical Chemistry

In this lab, you will test nineteen metallic ions **(cations)** with ten **reagents**. These cations represent members of several **"families"** from across the periodic table. The **Alkali, Alkali-Earth** families, and the 4th row **Transition** elements make up most of the samples to be tested. The ten reagents are solutions which are divided into **acids, bases, and salts.**

There are several purposes for this experiment, the first of which is to create a database of reaction properties of the known cations. These properties are unique to that element. The pattern of reactions can be used to identify an unknown sample. After testing the unknown in the same manner as the nineteen known materials, try to match the patterns. **If a reaction pattern of the unknown sample is identical to one of the known samples, then that is the identity of the unknown.**

Next, the reagents are classified for their activity. Some acids are very reactive, while others hardly react at all. The bases on the other hand are very reactive, as are the salts. The students are expected to create a priority list of which reagents to use first through last. In this way, the identity of the unknown will be determined in record time. *note: if you don't have Na$_2$S in the stock room, create some by reacting sulfur with a strong sodium hydroxide solution. Boil the mixture for 20 minutes. Cools and filter. The filtrate is sodium sulfide.*

Procedure:

.   Obtain ten clean test tubes.

.   Place six (6) drops of the reagents in the tubes, in the same order that they appear on the worksheet.

.   Place six (6) drops of the sample into each tube.

.   Shake the tube, and wait several minutes for the reaction to complete.

.   Write on the spreadsheet what has happened in each test tube.

.   Wash each tube with detergent and water.

.   Test all of the samples by repeating these first 6 steps.

.   Finally, test the unknown samples and identify them, based on their reactions.

Observation Code:

NR – no visible reaction

Solution – (Sol.) no visible particles formed, but some kind of color change in the solution

Suspension – (Susp.) tiny floating particles formed which **do not** settle out.

Colloid – (Coll.) thick, jelly-like particles which cling to the glass, create a film on the walls of the tube, and stick to each other, often forming a solid mass in the tube.

Precipitate – (ppt.) heavy, insoluble particles which easily settle out of solution.

	HCl	H$_2$SO$_4$	HNO$_3$	CH$_3$COOH	NaOH	NH$_4$OH	NaHCO$_3$	NaI	Na$_2$S	NaCl
Li										
Na										
K										
Mg										
Ca										
Sr										
Ba										
Cr										
Mn										
Fe										
Co										
Ni										
Cu										
Zn										
Al										
Sn										
Ag										
Hg										
Pb										

LAB #31

Formula of a precipitate (Advanced Placement version)

Purpose:

To investigate the theoretical yield of Lead Iodide and the formula of the compound. This lab requires the testing of eight combinations of the two reagent solutions. The purpose is to determine the actual **combining ratio** of the elements involved, and therefore the **oxidation value** (valence) of those reactors.

Diagram:

Draw or attach a photo of the equipment in set-up mode

Procedure:

Day #1

. In a 250 X 25 test tube, mix **5 ml of a 0.5 Molar solution of Lead Nitrate** with **40 ml of a 0.5 Molar Sodium Iodide solution**. label this sample #1

. Make the other seven combinations following the volumes in your data chart.

. Transfer to an ice bath while you set up the filtering apparatus.

. Cool solution on ice for at least five (5) minutes.

. Obtain eight pieces of #1 (slow) Whatman Filtering Paper and weigh it to four decimal places.

. Create eight paper cones with the filter paper and insert into the glass funnel of eight filtering stations. (four lab groups each doing two samples.

. Pour the mixtures onto the corresponding paper cone. The filtrate should be colorless and clear. The yellow lead iodide should be trapped on the paper.

. Wash several times with **chilled distilled** water to remove by-products and unreacted materials.

. Place filter paper into a drying oven overnight or lay flat on watch glass.

Day #2

. Reweigh filter paper (final mass) and calculate the actual yield. Compare to the theoretical yield.

DATA:

complete the spreadsheet attached

Graph:

Graph the expected vs the actual

Calculations:

* **Show the entire balanced reaction using a Mass-Mass relationship format.**

* Show all work in HSPT/HSPA format to complete the data table for each sample tube combination

Conclusion:

. The ***actual mass*** of the precipitate produced is rarely as high as the ***expected mass***.

. Propose a reason why this phenomenon is usually observed. (you can't get 100% yield) ***To research the reason, investigate the concepts of Ksp and collision theory.***

. This experiment required the use of cold water and chilled solutions to work properly. WHY? ***Solubility and temperature***

. What is indicated by the shape of the graph (actual <u>combining</u> ratio)

The highest yield matches the actual combining ratio.

. Explain the impact on the "limiting factor" on the expected mass.

AP Chemistry
molar solutions
% yield and limiting factor

Tube	$Pb(NO_3)_2$ Volume	NaI Volume	Ratio	Moles Pb	Moles I	mass Pb	mass I	Expected PPT	Actual PPT	% yield
1	5 ml	40 ml	1 to 8	0.0025	0.0200	0.518	0.635	1.1525		
2	10 ml	35 ml	1 to 3.5	0.005	0.0175	1.035	1.270	2.3050		
3	15 ml	30 ml	1 to 2	0.0075	0.015	1.553	1.905	3.4575		
4	20 ml	25 ml	4 to 5	0.0100	0.0125	1.29	1.588	2.8775		
5	25 ml	20 ml	5 to 4	0.0125	0.0100	1.035	1.270	2.3050		
6	30 ml	15 ml	2 to 1	0.015	0.0075	0.776	0.950	1.7263		
7	35 ml	10 ml	3.5 to 1	0.0175	0.0050	0.518	0.635	1.1525		
8	40 ml	5 ml	8 to 1	0.0200	0.0025	0.259	0.318	0.5763		

Ksp of Lead Iodide = 1.8×10^{-8}

concentration of each solution = 0.5 Moles/liter

Total volume in each tube = 45 ml

molar mass of Lead Iodide = 461 gm/mole

Limiting Factor -- % Yield -- Solubility

Tube	Pb(NO$_3$)$_2$ Volume	NaI Volume	Ratio	Moles Pb	Moles I	mass Pb	mass I	Expected PPT	Actual PPT	% yield
1	5 ml	40 ml	1 to 8	0.0025	0.0200	0.518	0.635	1.1525	0.8970	
2	10 ml	35 ml	1 to 3.5	0.005	0.0175	1.035	1.270	2.3050	1.6893	
3	15 ml	30 ml	1 to 2	0.0075	0.015	1.553	1.905	3.4575	3.2585	
4	20 ml	25 ml	4 to 5	0.0100	0.0125	1.29	1.588	2.8775	2.5536	
5	25 ml	20 ml	5 to 4	0.0125	0.0100	1.035	1.270	2.3050	2.0872	
6	30 ml	15 ml	2 to 1	0.015	0.0075	0.776	0.950	1.7263	1.3465	
7	35 ml	10 ml	3.5 to 1	0.0175	0.0050	0.518	0.635	1.1525	0.8990	
8	40 ml	5 ml	8 to 1	0.0200	0.0025	0.259	0.318	0.5763	0.4668	

Ksp of Lead Iodide = 1.8 X 10^{-8}

concentration of each solution = 0.5 Moles/liter

Total volume in each tube = 45 ml

molar mass of Lead Iodide = 461 gm/mole

Limiting Factor -- % Yield – Solubility

1.1525	0.897
2.305	1.7893
3.4575	3.2585
2.8775	2.5536
2.305	2.0872
1.7263	1.3465
1.1525	0.899
0.5763	0.4578

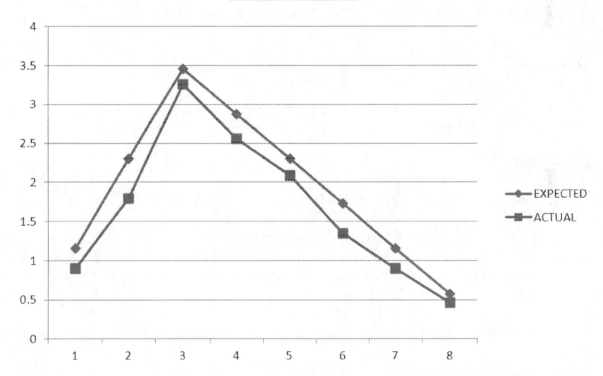

LAB #32

Laser Crystallography

Purpose:

using a Laser pointer, determine the atomic spacing is a crystal of Calcite

Diagram:

Draw or attach a photo of the equipment in set-up mode

Procedure:

. Measure a distance of 30 feet **(914.4 cm)** from the crystal sample to the wall

. Secure a sample of metal screening in a ring clamp

. Position a red laser pointer behind the screen so that the beam passes through the screen and is projected onto the wall 30 feet away.

. Accurately measure the spacing between the red dots on the wall. Record this distance

. Calculate the distance between the threads of the screen.

. Compare to the actual measurements taken of the screen.

. Repeat step 1-5 using a piece of Calcite, easily obtained from any school rock & mineral collection..

DATA:

	Screen	**calcite**
Distance from sample to wall	_914.4_ cm	_914.4_ cm
Wavelength of the red laser	_4100 _nm	_4100 _ nm
Distance between points of light	_____ cm	_____ cm
Calculated spacing	_____ nm	_____ nm
Actual spacing	_____ nm	NA

Calculations:

Atomic spacing = (beam length) * (point spacing) / wavelength

Beam length = 914.4 cm   actual spacing (screen) = 1 cm / # of wires

Wavelength Red laser = 4100 nm

Conclusion:

Write an essay about the usefulness of knowing atomic spacing. In reality, X-Rays are used in place of red Laser light. The wall projection screen is also replaced with either photographic paper or a CCD. Explain why this is necessary. **[hint: X-rays vs visible light at the dentist]**

LAB #33

Silly Slime

Materials:

Elmer's Glue (**Poly Vinyl Alcohol**)

Water

Borax Washing Powder (Sodium tetraBorate decaHydrate) $Na_2B_4O_7 * 10\ H_2O$

Mixing cup

Stirring rod

Directions:

. Place a quarter cup water into a cup

. Add a quarter cup of Elmer's White Glue to the water above.

. Stir mixture until all lumps are gone.

. Add a half cup of concentrated Borax solution. (**super saturated**)

. Stir until a gelatin forms.

. Remove the gelatin from the container and knead it with your wet hands until all the extra water is driven out   ***DO THIS OVER A SINK or BASIN***

. The more you knead the putty, the more elastic it becomes. This creates additional hydrogen bonding sites in the molecule, locking up more of the water.

LAB #34

Properties of Carbon Dioxide gas

Purpose:

Pure calcium carbonate (limestone) reacts with acids to liberate carbon dioxide gas. When the carbonate comes in contact with any acid, it forms a salt, water, and the gas. Carbon dioxide is slightly soluble in water. It also reacts with water to form a weak acid solution, called carbonic acid. In this lab, you will produce carbon dioxide and then test some of its properties.

Diagram:

Draw or attach a photo of the equipment in set-up mode

Procedure:

. Set up the gas generator as instructed.

. Fill the pneumatic trough with water until it flows out of the overflow hose.

. Place the carbonate (limestone) into the jar of the generator and cover with 20 ml of distilled water. This water will **moderate** the reaction and slow it down.

. Fill five large test tubes with water and place them standing inverted in the trough.

. Slowly pour the acid into the thistle tube of the generator. If the reaction stops before all the tubes are filled, add more acid.

. Allow the gas in the generator to flow for 30 seconds before collecting any gas in the test tubes. This will purge the generator of any air in the system. You want to collect a pure sample and not a mixture of CO_2 and air.

. Test each tube as instructed

Observations:

. What is the color of the carbon dioxide gas collected? _____

. Using a disposable pipet, dispense 2 ml of universal pH indicator into **tube #1,** filled with CO_2. Place a cork onto this tube. Shake vigorously. Record what happens.

. Light a wood splint and place it into **test tube #2.** Record what happens

. Hold an inverted **tube #3** over a candle flame. Record what happens.

. Turn **tube #4** right side up and quickly pour a few ml of calcium hydroxide solution into **tube #4.** Place a cork onto the tube. Shake the tube for several seconds and record what happens.

. Light a piece of pure magnesium over the candle flame. Quickly place it into **tube #5** and record what happens.

Conclusion:

Using class notes, your text, and the internet, prepare a list of properties for carbon dioxide gas. Compare this list to your observations.

LAB #35

Properties of pure Hydrogen gas

Purpose:

Pure calcium metal is very reactive with water. When the metal comes in contact with water, a strong base is produced and hydrogen gas is liberated. In this lab, the hydrogen will be collected and then tested for its properties. The most notorious property of hydrogen gas is that it is explosive. Due to the high reactivity of calcium, zinc metal will be used in this experiment to generate the hydrogen gas.

Diagram:

Draw or attach a photo of the equipment in set-up mode

Procedure:

. Set up the gas generator as instructed.

. Fill the pneumatic trough with water until it flows out of the overflow hose.

. Place a small piece of **zinc metal** into the jar of the generator.

. Fill several large test tubes with water and place them standing inverted in the trough.

. Pour 100 ml of 3.0 Molar HCl solution into the thistle tube of the generator.

. Collect a tube of gas immediately after starting the generator. **(TUBE #1)**

. Allow the gas in the generator to flow for 30 seconds before collecting additional gas in the test tubes.

. test each tube as instructed

Observations:

.   What is the color of the hydrogen gas collected? _____

.   What is the behavior of the gas as tested (#2) with wet litmus paper?

Red litmus test _____ Blue litmus test _____

.   Light a wood splint. Place it near the mouth of a **test tube #1**. What happens? Perform this same test on Tube #3. What happens.

.   Collect one tube that has been half filled with an air bubble before the hydrogen was collected. How does it behave with the lit match? Compare to tube #1 & #3.

Conclusion:

Hydrogen gas is lighter than air. It is very flammable. It has certain physical properties. Research Hydrogen's properties on the internet and write a fact filled essay about the history, commercial production, and properties of hydrogen gas. Give as much information as you can find. Use more than one source.

Research the "Hydrogen Highway".

LAB #36

Properties of pure Oxygen gas

Purpose:

Potassium chlorate will decompose, if heated, into potassium chlorise salt and pure oxygen gas. This reaction can be dangerous! To make the decomposition safer, a liberal quantity of Manganese Dioxide is mixed with the $KClO_3$ to slow the reaction. The MnO_2 is therefore a catalyst. The mixtures should be 3 parts chlorate and 1 part dioxide.

Diagram:

Draw or attach a photo of the equipment in set-up mode

Procedure:

. Set up the gas generator as instructed.

. Fill the pneumatic trough with water until it flows out of the overflow hose.

. Place a quantity of the $KClO_3$ and MnO mixture into a large test tube. Use a amount equal to the thickness of two fingers.

. Fill five large test tubes with water and place them standing inverted in the trough.

. Slowly heat the tube containing the chlorate mixture.

. Allow the gas in the generator to flow for 30 seconds before collecting any gas in the test tubes. This will purge the generator of any air in the system. You want to collect a pure sample and not a mixture of O_2 and air.

. Test each tube as instructed

Observations:

. What is the color of the oxygen gas collected? _____

. Using a disposable pipet, dispense 2 ml of universal pH indicator into **tube #1,** filled with O_2. Place a cork onto this tube. Shake vigorously. Record what happens.

. Light a wood splint then blow it out. Now place to glowing wood splint it into **test tube #2.** Record what happens

. Hold an inverted **tube #3** over a candle flame. Record what happens.

. Turn **tube #4** right side up and quickly pour a few ml of calcium hydroxide solution into **tube #4.** Place a cork onto the tube. Shake the tube for several seconds and record what happens.

. Light a piece of pure magnesium over the candle flame. Quickly place it into **tube #5** and record what happens.

Conclusion:

Using class notes, your text, and the internet, and create a list of properties for oxygen gas. Compare this list to your observations.